Barndominium
Floor Plans and Designs

For Beginners

A Complete Guide to Creative Ideas, Functional Designs, and Custom Layouts for Building the Perfect Modern, Spacious, Customizable Home

Cassius Peregrine

Disclaimer and Terms of Use

The author and publisher of this book and the accompanying materials have used their best efforts in preparing this book. The author and publisher make no representation or warranties with respect to the accuracy, applicability, fitness, or completeness of the contents of this book. The information contained in this book is strictly for informational purposes. Therefore, if you wish to apply the ideas contained in this book, you are taking full responsibility for your actions.

Printed in the United States of America

TABLE OF CONTENTS

TABLE OF CONTENTS III

INTRODUCTION 1

CHAPTER ONE 2

WHAT IS A BARNDOMINIUM? 2

    Understanding Barndominium floor plans 4

    Is a bardominium cheaper to build than a house? 6

        The general breakdown of common barndominium floor plans. 8

CHAPTER TWO 16

IMPORTANT FACTORS FOR BARNDOMINIUM FLOOR PLANS 16

    Understanding Barndominium Layouts 21

        Determining Your Family's Needs 21

        Space allocation: key decisions for every room 26

        How to Future-Proof Your Barndominium 27

        Budget: How your floor plan affects the overall cost 27

        How to choose the correct land size and location for your barndominium 36

CHAPTER THREE 39

COMMON FEATURES IN BARNDOMINIUM FLOOR PLANS 39

        Open Floor Barndominium Designs 43

    Flexible Barndominium Floor Plan Design 44

        Flexible Living Revolving Around Multi-Purpose Rooms 47

        Approaches to Design that Achieve Maximum Versatility 48

        Pros of Lofts and multi-purpose rooms. 49

        Attributes that Make Outdoor Living Spaces Attractive 51

        Wraparound Porches: Design Considerations 52

        Considerations to Make in Patio Designs 53

        Why You May Need Extra Garage Space 54

        Changing the Appearance of Your Garage and Workshop 55

        Multi-Use is Not Constricted to Traditional Garage 55

        Popular Barndominium Floor Plans for Workshops and Garages 56

CHAPTER FOUR 58

TYPES OF BARNDOMINIUM FLOOR PLANS 58

    Small: 1,000-1,500 sq. ft; 1-2 bedrooms 58

CHAPTER FIVE 78

MAKING YOUR BARNDOMINIUM FLOOR PLAN FUNCTIONAL 78

How Kitchen Islands Factor in with the Design Process 85

Key points in the En-Suite Bathroom 88

Design Elements of a Walk-in Closet 89

Characteristics of an Outdoor Private Space 90

Decks: How They Reinvent Outdoor Living 91

Incorporating outdoor living spaces into the floor plan of a barndominium 93

CHAPTER SIX 94

DESIGNING FOR EFFICIENCY AND COMFORT 94

How Energy-Efficient Appliances can Lower Your Electricity Bill 101

Strategies for Incorporating Energy Efficiency in Floor Plans for Barndominium Designs...................... 102

Importance of Ventilation/ Airflow 102

Practical Instructions for Better Air Circulation in Design 104

CHAPTER SEVEN 112

CHOICES REGARDING FLOOR PLANS 112

Two-Story Plans: Save space by pushing private spaces upstairs 115

Advantages of Moving Private Spaces on the Upper Ground Floor 116

Standard Two-story barndominium Floor Plans House Features.. 117

How to Customize a Two-Story Barndominium 118

Knowing the Difference between L-shaped and U-shaped layouts.. 119

Improved Flow between Indoors and Outdoors... 120

The courtyard designs: The heart of L-shaped and U-shaped homes. 121

Making the most of the views and accessing nature 123

CHAPTER EIGHT 124

PRACTICAL SPACES IN BARNDOMINIUMS 124

The Attached Garages and Workshops... 124

The Detached Garages and Workshops 126

Emerging demand for automobile and recreational vehicle storage 132

Different types of automobile and recreational vehicle storage options... 132

Advantages of Using Vehicle and RV Storage 134

Things to Consider When Choosing a Storage Facility 135

CHAPTER NINE 136

ACCESSIBILITY CONSIDERATIONS 136

The importance of wide hallways in navigable ways.. 142

Low countertops with an accessible kitchen setup 142

Eliminate entry barriers with no-step entries.. 144

Bathroom accessibility: ensuring individual safety and comfort 145

CHAPTER TEN... 148

SAMPLE BARNDOMINIUM FLOOR PLANS **148**

VISUAL EXAMPLES AND BLUEPRINTS OF POPULAR DESIGNS 148

    Case studies in different layouts *157*

    Pros and cons of various designs *159*

CHAPTER ELEVEN **163**

TIPS FOR CHOOSING OR CREATING A BARNDOMINIUM FLOOR PLAN **163**

WORKING WITH PROFESSIONALS: ARCHITECTS AND DESIGNERS 163

    Off-the-shelf Plans Versus Custom Plans: The Positives and Negatives of Each *166*

BARNDOMINIUM OFF-THE-SHELF FLOOR PLANS 167

    Custom Barndominium Floor Plans *169*

CHANGING PLANS: HOW YOU CAN TAKE ANY DESIGN AND MAKE IT YOUR OWN 174

CONCLUSION **179**

INTRODUCTION

Barndominiums are a unique fusion of barn and condominium that has recently gained momentum among owners who design and build personalized, reasonably priced dream homes. Originating from functional barns with living spaces, these have transformed into multi-functional and affordable homes. Contemporary interpretations of rural living, barndominiums most especially thrive in rural settings or on large plots of land right outside urban areas. The major advantage they boast is that they are cheap and affordable when compared to conventional homes; this enables homeowners to consider building barndominiums while still making full use of the facilities of modern life.

One of the major selling aspects of a barndominium is the toughness in which the design comes, especially when steel is integrated into it. When compared to conventional homes, metal barndominiums are more resistant to many of the common problems that destroy them, such as bugs, water, and fire. Besides requiring low maintenance, these steel-construction homes can be more sustainable, since they tend to last longer with fewer repairs. Their level of durability is particularly helpful for those living in areas prone to extreme weather, providing both peace of mind and a long-term savings advantage in terms of repairs and upkeep.

Barndominiums provide one with loads of leeway when it comes to floor plans. Living spaces can be extended in many ways other than conventional rooms. From workshops to RV garages, bonus rooms, and pole barns-even horse stalls or airplane hangars, a barndominium can be tailored for very particular needs in your lifestyle. With big open floor plans and your choice of rustic, modern, or industrial building, the design possibilities are virtually endless.

Arguably, the most effective way to build a barndominium is by using a steel building kit, provided by companies like Worldwide Steel. This gives one an extremely practical and cost-effective method of building a robust building while being able to specify every little detail. The strength of steel mixed with the versatility of the barndominium floor plan finally allows homeowners to build a space that is both functional and completely reflective of their style and needs.

CHAPTER ONE

What is a Barndominium?

Barndominiums popularity is on the rise. Barndominiums has taken off inside the housing market within the last ten years. Once considered a somewhat niche style in rural areas, it is now a full-fledged mainstream phenomenon. In its essence, a barndominium merges the traditional functional features of a barn with the comfort and functionality of a modern home. This unique architectural concept offers wide-open spaces, affordability, and customization capabilities highly in demand by homeowners, especially those looking for alternatives to conventional home construction.

Origins and Evolution

The origin of the barndominiums was simple barn conversions. In most cases, farmers and ranchers look to save costs on construction with the simple thought of building a barn with some living quarters for farm workers or putting a family of their own. These structures were totally practical and cost-effective to have, usually with metal exteriors and open floor plans.

Soon, over time, it made lots of sense for homeowners to want to convert the barn into a home. As people looked for alternatives other than typical conventional houses, the rustic charm with versatility of the barndominium made them an ever-appealing sight. The style did start to bring in more modern conveniences and design elements, coming into being structures that are a mix of industrial functionality and homey warmth.

A Unique Appeal

The growing popularity of barndominiums is for a variety of reasons. The main ones include cost. Traditional home construction has grown very expensive due to increased material costs, labor shortages, and zoning regulations. Most barndominiums, particularly those with metal frames, are more affordable to build. Their simple construction process often means that, for many, homeowners can either self-build or use smaller, local contractors, further driving down costs.

Another great thing about barndominiums is the flexibility in their design. From a minimalist loft style to an incredibly elaborate country retreat, these buildings can offer a great degree of personalization. Many opt to incorporate expansive open floor plans, high ceilings, and large windows that allow abundant natural light and give a sense of spaciousness. This allows the barndominium to be an exciting choice for families, retirees, or any one person seeking a very personalized house.

Moreover, strength and low maintenance make them more popular. Among these, metal-framed barndominiums can bear bad weather conditions quite effectively compared to conventional houses. Therefore, they are fully suitable for areas with frequent storms, fires, and other natural catastrophes. Not least in importance, most of them save energy since they mostly require less heating and cooling, allowing the homeowners to reduce the costs of utility bills over a certain period.

The Rural Lifestyle and Urban Interest

Barndominiums first gained widespread acceptance in rural areas, especially in the south. For agricultural or ranching communities, it made much sense that the owners should live in an area that would also house tools and equipment or livestock. Today, the style of living epitomized by barndominiums-a mix of rustic charm with modern living-has found its appeal to urban dwellers and many desiring to break free from suburban life.

They want to live on larger plots of land but also want a house that combines simplicity and a rural lifestyle with modern facilities. The work-from-home revolution has only put gasoline on this curiosity, with many professionals now looking to get away from bustling cities for space and calmness. Barndominiums offer these a solution unlike any other: affordable, capacious homes to be erected on expansive properties.

Eco-Friendly Living

The whole barndominium trend has also been commanding attention from eco-sensitive homeowners. Many barndos are built with materials either recycled or sustainable. And with energy-efficient designs combined with solar panels, rainwater collection, and other green technologies, they do make for a very attractive option for many.

Barndominiums, to a great extent, embody the very notion of minimalism and eco-friendliness. First, their open-design floor plans reduce the need for using massive amounts of building materials. Second, their reduced environmental footprint manages to fall in step with the growing trend toward sustainable homebuilding.

Understanding Barndominium floor plans

These include everything from very flexible floor plans to very expansive ones. Some of the exteriors tend to be a bit more uniform; however, they can still be pretty striking. It's truly inside these homes where the variation begins: some of the interior spaces are owners who choose sophisticated layouts-large bedrooms, and large bathrooms-while others prefer the simplicity of an open-concept design.

The leading factor for most is the number of bedrooms. Since barndominiums consist of a metal shell, their internal configuration can be very flexible. While typical plans range from 4,000-square-foot homes with 2.5 bathrooms (2 full bathrooms and a half-bath) and three bedrooms, it's not out of the ordinary to find a few designs ranging in size from two to four bedrooms.

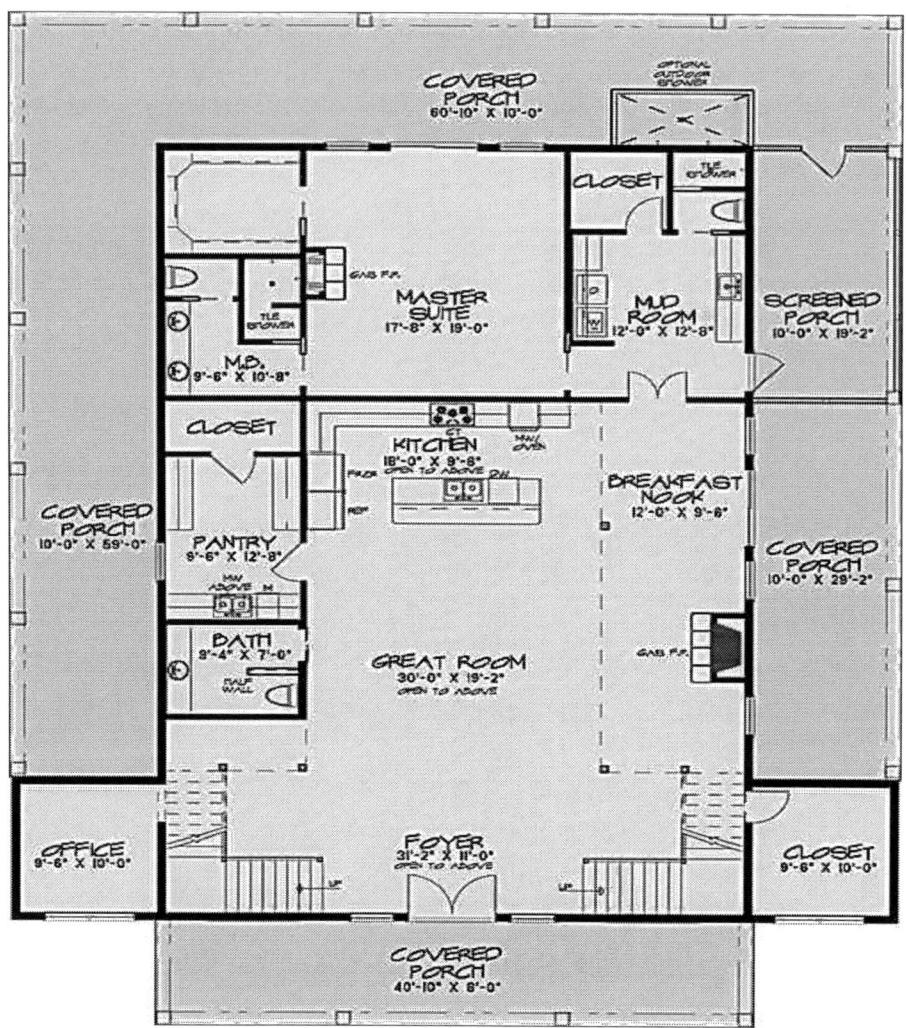

They aren't confined to small sizes either. While many of them are modest in size, there are a lot of big ones, too. Since they have historically been used as agricultural buildings with immense space to store farm equipment and do work, barndominiums can be as spacious as mansions. Those who focus on open floor plans will find many that rival the size of traditional large homes.

For singles, the studio layout may work out best, or for the growing family, expanded floor plans can be a direct consideration. The barndominiums offer an addition over time option as well, although one needs to consider the shape and size of the original structure. Most plans are designed around the metal frame foundation and may affect some leading design factors.

Barndominiums come in a variety of sizes but are often rectangular or square. Porches and/or patios are extra features outdoors that seem to add the finishing touches, though oftentimes homeowners complete the view with their taste in strategically placed windows.

You can find barndominiums in sizes such as 30×20 feet, 40×30 feet, 40×60 feet, 50×75 feet, and 80×100 feet floor plans. The larger-sized barndominiums, like those that are 80×100 feet, will usually have four bedrooms and at least two bathrooms, along with extras, such as utility rooms, mud rooms, walk-in closets, shops, pools, and even mother-in-law suites. Most of these spacious homes will have living rooms that measure 30×25 feet or greater.

Smaller barndominiums are a bit different, a structure of 20×30 feet could contain one or two bedrooms, one laundry, a kitchen, a family room, a bathroom, and a closet. In these smaller designs, the family room is usually around 12×12 feet.

Is a bardominium cheaper to build than a house?

The increasing fascination with Barndominiums is found in modern times as many people move away from traditional homes into them, and for good reasons, too: some advantages that it offer to family living and everyday life.

Among the many great advantages, perhaps the most outstanding thing about barndominiums is that they are much more cost-effective. Other homes tend to be fairly expensive, whereas families can often buy a lot more and get their money's worth when it comes to barndos. Finding barn-style homes up to 50 percent cheaper than their conventional counterparts is not at all out of order on the market.

Another driving force in the popularity of barndominium homes is durability. Made out of solid metal, barndominium houses share most of the same qualities as extremely durable steel buildings. Barndos are resistant to fire, mold, harsh weather conditions, infestation by termites, rot, and mildew. If you have concerns about the damage of a house due to the weather or structural integrity over time, consider a barndominium.

Building a barndominium is also quite uncomplicated. The metal roofing systems and siding are very easy to install, and the whole construction tends to move fast, so it is pretty convenient and efficient for the construction of barndominiums.

If you're an eco-conscious person or try to save on monthly utility bills, then a barndominium will come as an excellent option for you. Their metal construction makes them well-insulated, which may help with energy efficiency in heating and cooling homes.

When it comes to maintenance, which is the real ace in the hole of the barndominium, the amount required is minimal since they were designed for agricultural use. For people with heavy schedules, this can be an awesome boon to flexibility and free time to focus on what is most important.

Additional reasons the flexibility of barndominiums goes beyond maintenance; they are easy to alter or extend-whether one wants to add a new wing, install a swimming pool, or make other adjustments, and customization is easy and practical.

Above all, barndominiums are very flexible. They can be a main house but can also be used for other functional purposes such as a woodshop, guest house, pool house, game room, garage, office space, and even a clubhouse area. Whatever the need is, a barndominium can go according to your style.

The appeal: inexpensive, durable, and easy to customize

The barndominiums are the ultimate combination of affordability, strength, and adaptability. They started largely as agricultural structures but transformed into modern living spaces to fit any need in lifestyles. From building a home or workshop to a business, barndominium floor plans have big benefits for homeowners and builders alike.

Affordable to Build

Perhaps the biggest selling point for most barndominium floor plans is their affordability. Constructing a conventional house is an expensive affair; barndominiums present the possibility of more reasonable rates. Since these are often designed with either a metal frame or steel exteriors, they tend to be cheaper to build compared to conventional brick-and-mortar homes. Their simple structure helps reduce the actual material and labor costs, thereby accelerating the construction process and making it more affordable. Another component driving down the cost is the expansive plots of land on which they can be built, thereby steering clear of many expenses related to expensive urban property. Many of the barndominiums are derived from pre-fabricated kits, which further reduces the cost and leads to faster construction.

Strength and Durability

Barndominiums are also characterized by their high levels of strength. Since they are usually constructed from steel or metal materials, these houses are resistant to many common problems related to traditionally wood-framed homes: termites, rot, and warping. They are highly resistant to extreme weather conditions that include heavy winds, hail, and even fire. The metal constructions will not only guarantee longevity but also reduce long-term maintenance costs. They are designed to last for decades with very low maintenance; hence, barndominiums represent a very good investment for people in search of a low-maintenance yet durable home.

Easy Customization

Another appealing thing about a barndominium floor plan is its flexibility in design and customization. The open-concept layout is generally a signature of barndominiums and allows enough room for creative interior design. Because the structure does not usually depend on load-bearing interior walls, there is no restriction to homeowners in customizing the layout to whatever suits their purpose. You may want an expansive open living area, multiple bedrooms, or an attached workshop; barndominiums can serve all these purposes with a wide range of floor plans. Large windows, wraparound porches, and high ceilings evoke comfort and aesthetic feelings in modern designs.

In the end, barndominiums are an attractive alternative to traditional houses, offering in their structure a combination of affordability, durability, and flexibility. Barndominiums are ideal for those looking for an affordable yet durable and highly adaptable living space.

The general breakdown of common barndominium floor plans.

Conceived as a combination of barns with living quarters, barndominium has grown into one of the most popular ways to provide modern amenities within a rustic setting. From open-concept space for family living to multi-purpose layouts complete with workspaces, there is a lot of design leeway for barndominiums. Here's a general breakdown of some of the most common barndominium floor plans:

1.   **Open-Concept Layouts**

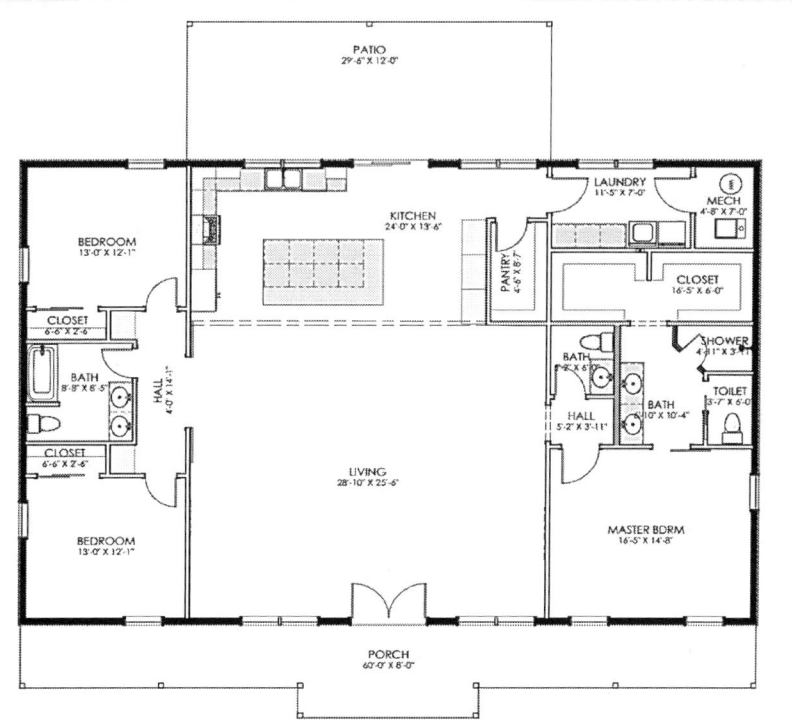

One of the signature staples of many barndominium floor plans is the open-concept design. That usually means large, un-partitioned spaces that encompass everything from the kitchen and dining areas to a living area all in one large room. The lack of walls creates a space that feels open and airy, perfect for families who love to entertain or who simply want to feel connected within the home.

Open-concept designs center the living space within most floor plans, placing the kitchen and dining areas either adjacent or part of the same space. The ultimate in terms of ease of movement and communication would be an absence of barriers from room to room. Large windows and high ceilings, which are part of the general appeal of barndominiums, enhance an openness or lightness feeling.

2. Split-bedroom layout

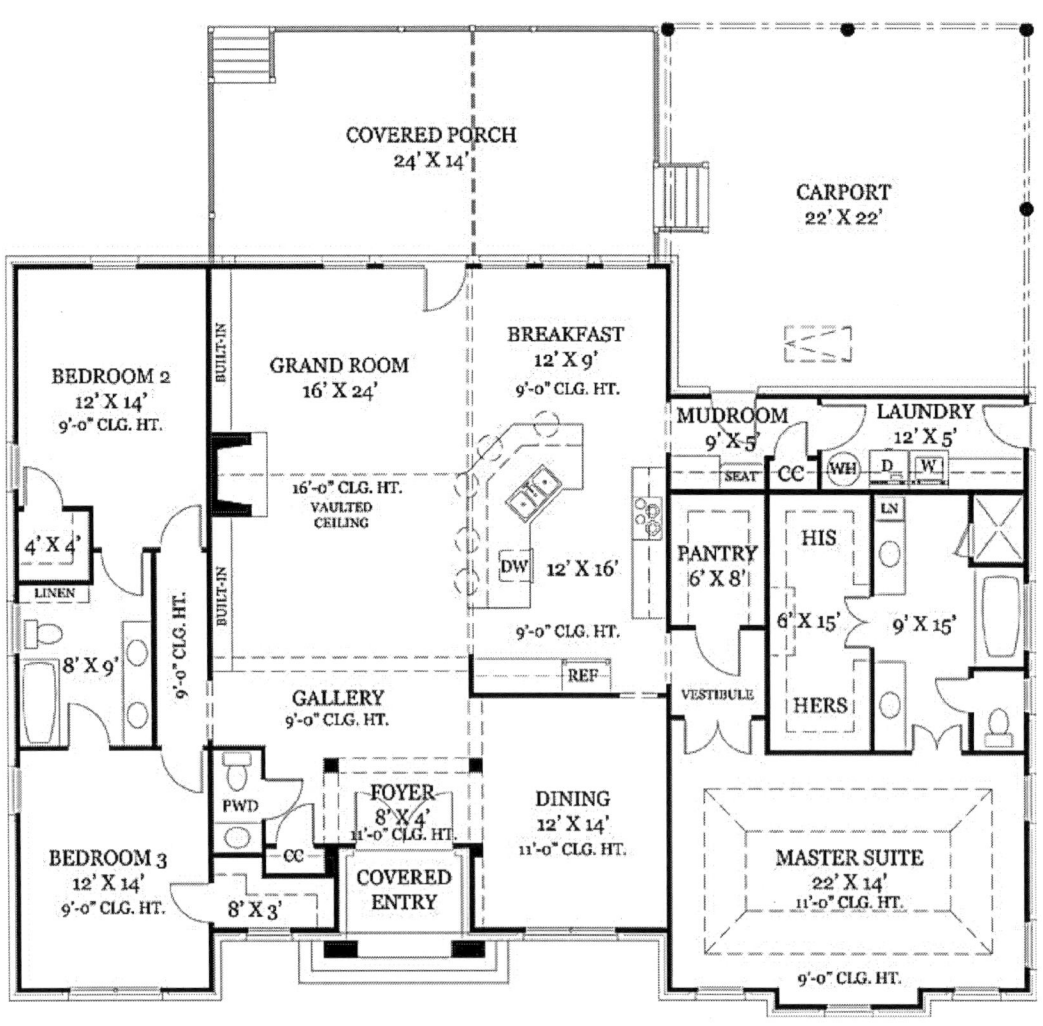

Another common design is that of a split-bedroom arrangement, having the master suite on one side of the barndominium and the other bedrooms on the opposite side. This works great for privacy, allowing the master suite to be in its enclosed area of space many times, with an attached bathroom and walk-in closet.

This layout is such that, in most cases, the central living area becomes a shield between the two sleeping zones. It works for families with children, or even multi-generational, with everyone getting their own space while maintaining flow within the home.

3. **Loft Space Designs**

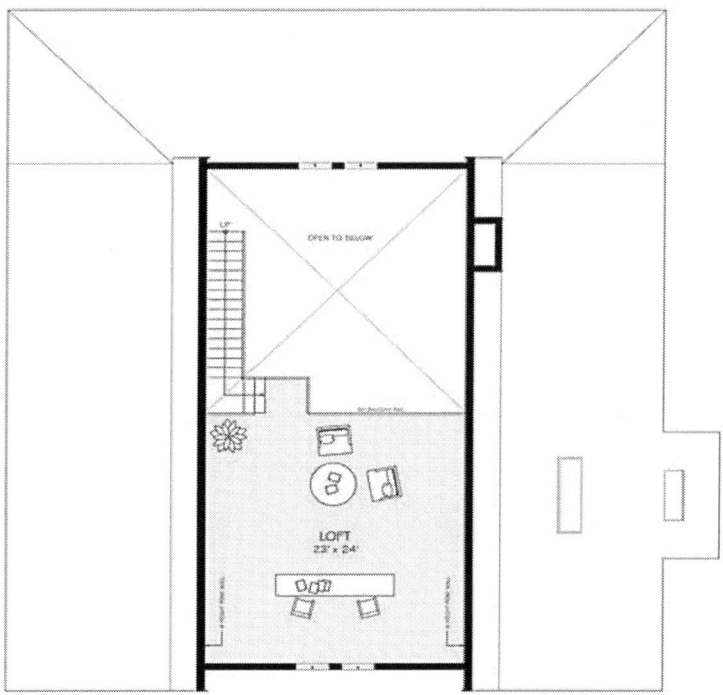

Barndominiums very often have loft spaces if the ceilings are high. Lofts add a second level to what would otherwise be a single-story house. This opens up more square footage without increasing the footprint. Second levels like this can be used for extra bedrooms, a home office, or even entertainment space, which allows flexibility with the functionality of the home.

Lofts also add visual interest because the staircase itself often is a design focal point. Also, the loft-over-looking main living space or tucked away provides an area that can serve a variety of functions in a cozy and secluded manner.

4. Combination Living and Workspace Floor Plans

One of the original uses for a barndominium was to serve both living and working purposes under one roof. So many modern floor plans still grasp that principle, incorporating workshop areas, garages, or even office spaces in the same building.

These floor plans may feature a large garage or workshop space attached to the living space, ideal for the hobbyist, mechanic, or anyone who requires a dedicated workspace. The key word here is versatility—barndominiums offer the ability to combine work, hobbies, and home life in one building.

5. U-Shaped or L-Shaped Floor Plans

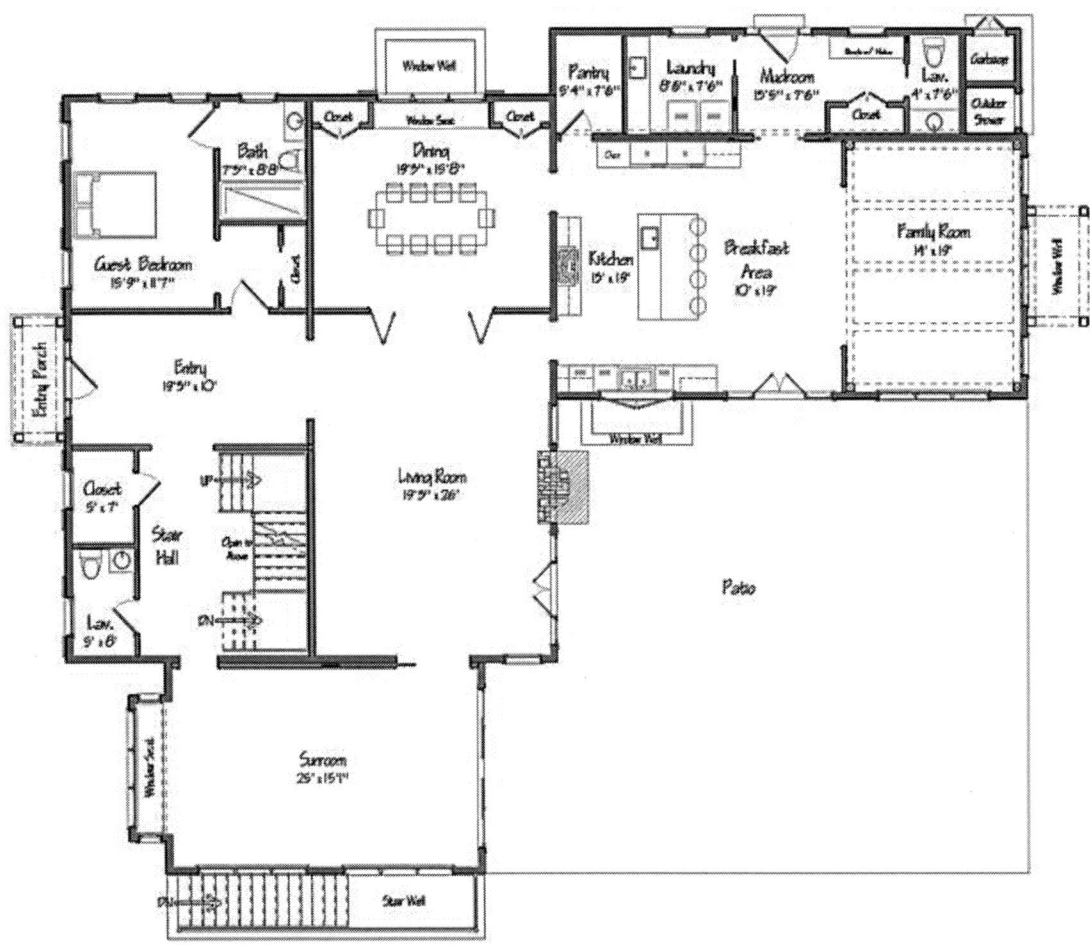

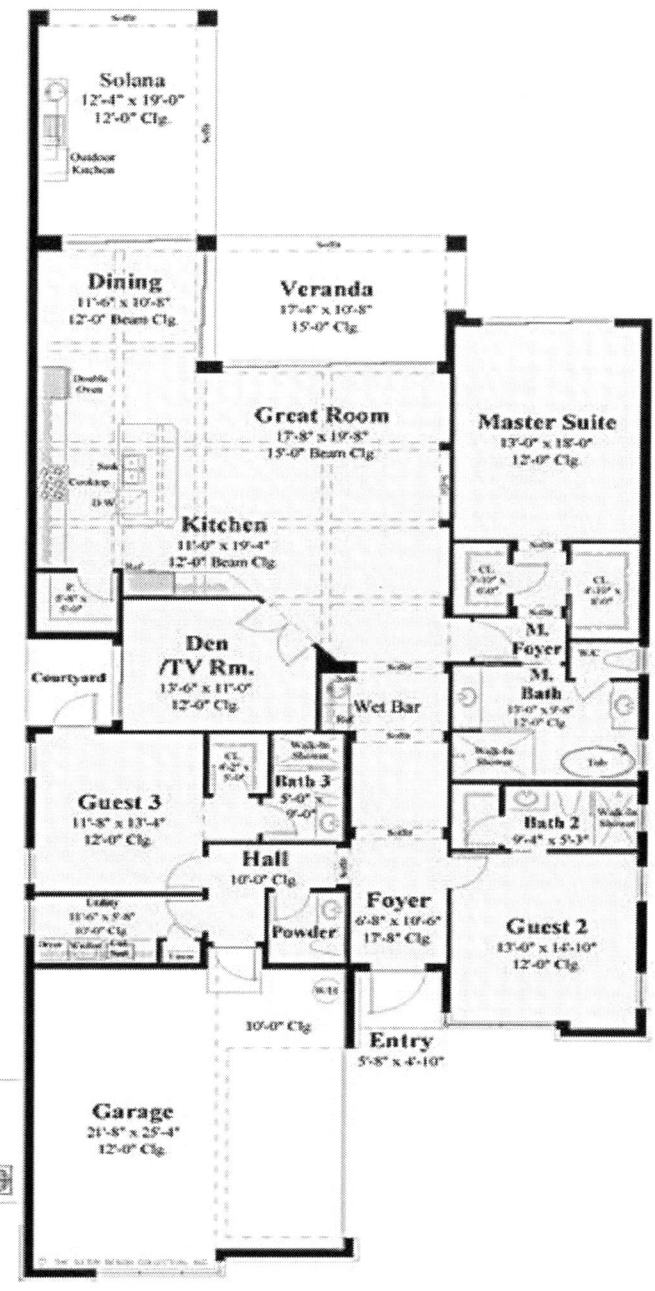

You will also see U-shaped and L-shaped designs fairly frequently in some barndominiums; these provide a more enclosed structure. Such layouts sometimes have a courtyard or outdoor living area in the middle or along one side of the residence. This could be ideal for those who enjoy the outdoors, as it would afford a private, somewhat protected spot for al fresco dining, entertaining, or even just a little gardening.

In a U-shaped layout, the bottom of the "U" generally contains the main living spaces, while the wings contain bedrooms and other private rooms. The L-shaped plan includes the common areas in one portion of the house, whereas the other portion contains the private ones, such as the bedrooms and offices. These will do great for maximizing available land space while keeping privacy and functionality intact.

6. Two-Story Barndominiums

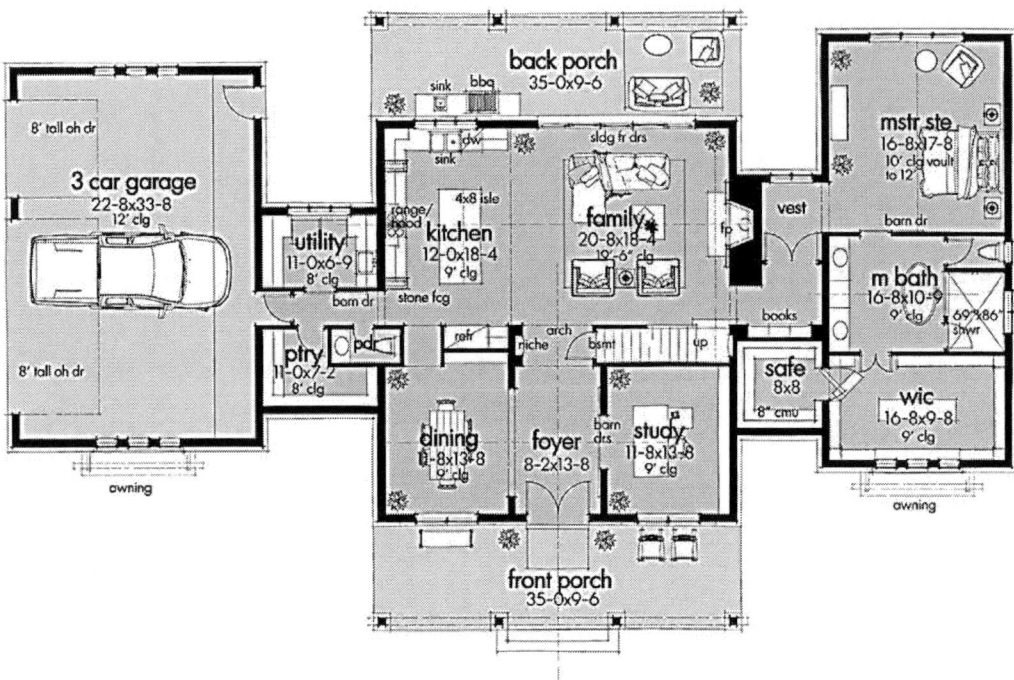

While many barndominiums are single-story, two-story options are popular. These floor plans allow owners to maximize their vertical space and be able to have more room without needing a larger footprint. In general, the first floor of a two-story barndominium would typically house the living spaces like the kitchen, living room, and dining room, with bedrooms and additional bathrooms on the second floor.

If this level is designed as an open area or balcony, it may overlook or extend to provide a view of the lower level, thereby maximizing the building's height. This would be great for families needing more square footage but prefer smaller lot sizes since the house can get more living space without expanding the ground-level footprint.

7. Multi-Function Rooms

Many barndominiums consist of multi-functional rooms designed with the idea of serving several functions. In other words, one huge room may be used as a family room, game room, or exercise room. Such rooms are incorporated into floor plans in cases when flexibility is needed, allowing one to easily make certain changes if the need arises over time.

Spaces such as this large and flexible are handier, surely, for a growing family or one that entertains frequently. Perhaps one of the biggest advantages of living in a barndominium is being able to switch the rooms around to use them for different purposes.

8. Wraparound Porches

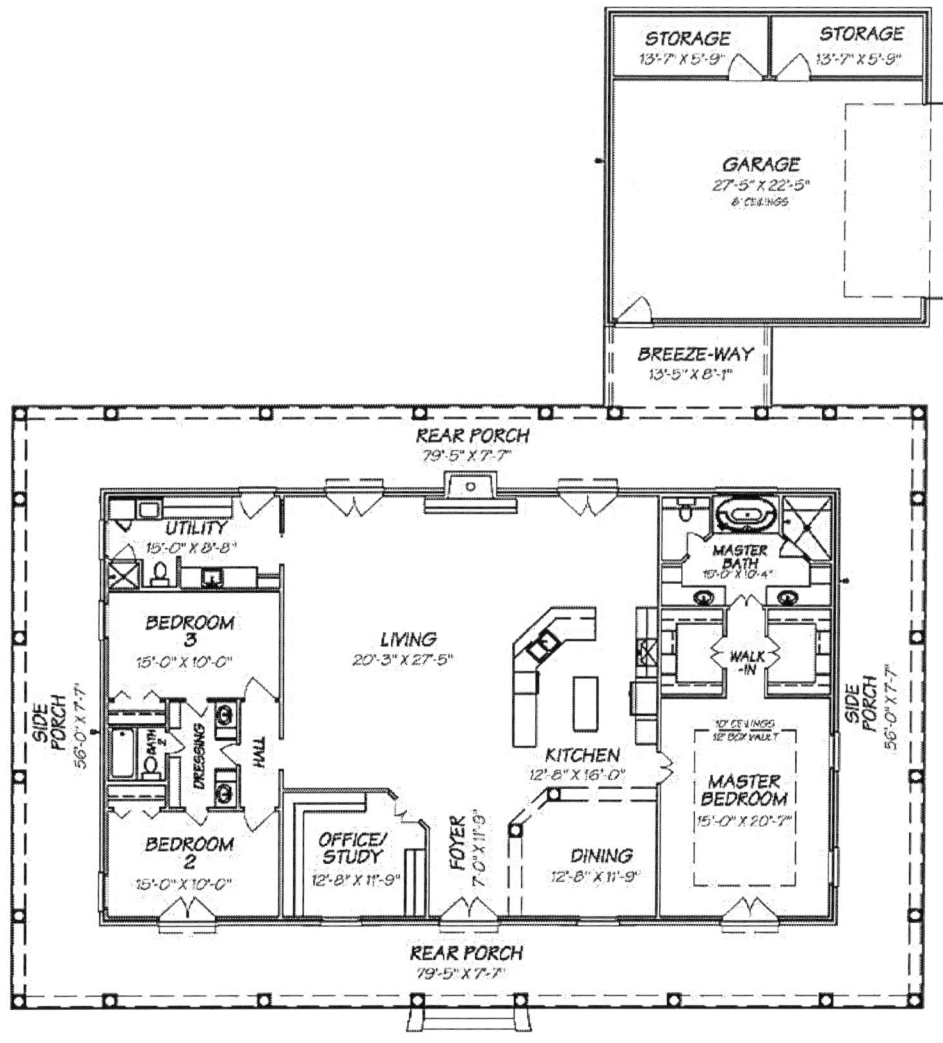

So many barndominiums take their cue from traditional barns or country homes and incorporate wraparound porches into their design. These extend one's living space outdoors, often being used to simply relax and enjoy the beauty of one's surroundings. In the designs for those floor plans with this option, the porch is often extended around the perimeter of the home, making it accessible from many different rooms.

CHAPTER TWO

Important Factors for Barndominium Floor Plans

To create a floor plan for a barndominium, one of the most important things to keep in mind is utility and flow. The barndominium is often used as both a living space and a workshop or barn; therefore, an obvious and sensible layout must be created. Facilities such as the kitchen, living, and dining areas should be located next to each other inside a house for more space and easy flow. Private places such as bedrooms and bathrooms have to be strategically situated to maintain privacy. For this reason, the transition from living to working rooms should be as smooth as possible without any noise or other inconveniences.

Besides these factors, natural lighting and ventilation are also very significant. Since barndominiums usually possess high ceilings and broad openings, it feels superb to add a large number of windows and skylights to allow maximum natural light into the room. This aside from reducing the amount of artificial lighting that might be required during the day, also helps in energy consumption reduction. Moreover, since most barndominiums are situated in rural areas, improving natural ventilation by carefully placing windows and ventilation systems will make the room cooler during summer and reduce the need for air conditioning.

Also, customization possibility and structural support are two of other key factors. In the case of great open areas without interior load-bearing walls, barndominiums are more often made with steel or metal frames that provide great flexibility. This allows a wide range of alternatives for fitting the interior design to your requirements, whether you are looking to create a more spacious workspace or a more intimate living space. Consultation with structural engineers is equally important because the owners might have intended to add more floors or make some modifications to the building for long-term durability and safety.

Purpose and Lifestyle: Full-time home or vacation getaway?

The custom design of any given barndominium floor plan is intended to suit numerous living circumstances, hence being both a permanent residence and a vacation retreat. Several reasons explain its recent rise in popularity: its versatility, cost-effectiveness, and the way rustic elegance meets modern convenience.

Purpose and Flexibility Applied

Perhaps the most striking feature about barndominiums has to do with the versatility that is associated with them. Although commonly associated with country living, the modern variety is equally practical for suburban and metropolitan living. The choice between a vacation getaway and a primary residence often depends on an individual's preferred lifestyle and long-term goals.

1. **Full-time dwelling:**

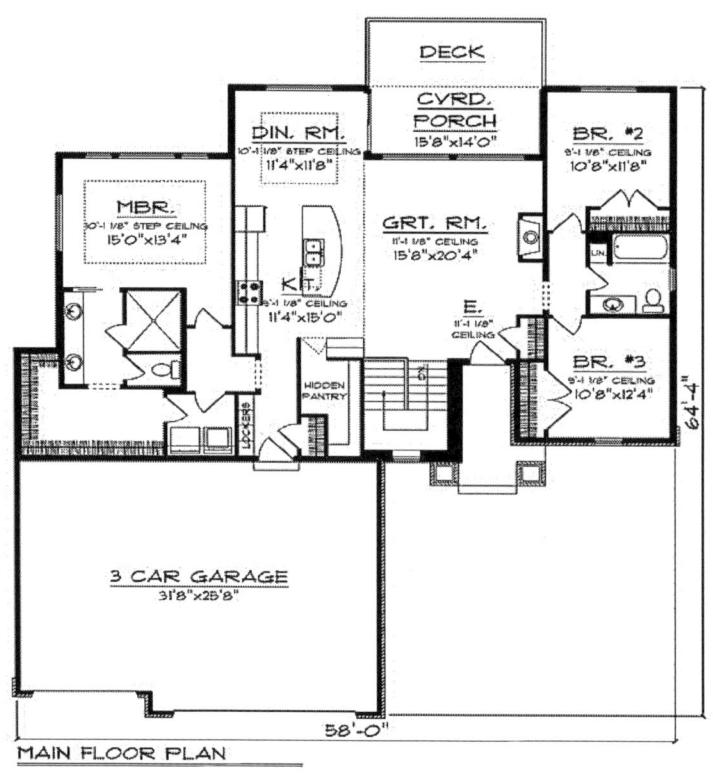

A barndominium floor plan can be used as a permanent dwelling for individuals, couples, or families who place a high emphasis on open spaces, simple designs, and lower building expenses.

Barndominiums have become renowned for their open and wide floor plans, which make them the perfect building option for people who value shared living areas highly. Speaking of entertainment, most barndominiums have a large, open space that is just great for company.

Flexible floor plans may also be personalized to suit either the needs of expanding families or the peace and spaciousness that an individual may require. Besides that, the steel-frame structure of most barndominiums offers long-term durability, and therefore, can be maintained at minimal levels, which is outstanding advantage for those home seekers wanting a house that is concrete yet low in its maintenance demand.

2. Vacation gateway

Being expansive in their living quarters, barndominiums would also do quite well for vacation retreats. Those who want to get away from the hustle and bustle of city life for a quiet, nature-oriented getaway may find a barndominium built in a rural or picturesque area to be their solution.

The interiors of these homes are big enough for you to hold huge gatherings, hence making them an ideal place for family reunions or friends' vacations. Equally, barndominiums work great for rentals, being able to also incorporate many bedrooms and open living spaces.

They offer a unique experience for stays, especially when vacationers want to get away into something different from the traditional cabins or vacation houses. For tourists who want to feel luxurious yet peaceful, rustic-style cottages amalgamated with modern amenities might be appealing.

Floor Plan Changes and Personalization Possibilities

Some of the most distinguishing characteristics of the barndominium include an open-concept layout, which allows for a broad variety of customization possibilities to be implemented. One can modify floor plans according to requirements for a particular lifestyle, whether permanent residence or holiday house.

1. Full-time living customization

The flexibility in the floor plans allows those families who would want to use the barndominium as their permanent dwelling to come up with a functional and highly individualized design. Large kitchens with huge islands are perfect for the family and great for family dinners and gatherings.

This opens up the space by incorporating natural light due to its design with high ceilings and large windows. Since barndominiums can be for both small and large families, they can include some bathroom and bedroom options.

Also, the owners can plan space for their hobbies or business by building offices, workshops, or house gyms onto their structures. Community living together with privacy can be achieved in the house by allowing one to create open yet discreet zones. This is suitable for those kinds of families that have several activities and needs.

2. Barndominium floor plans for your vacation house

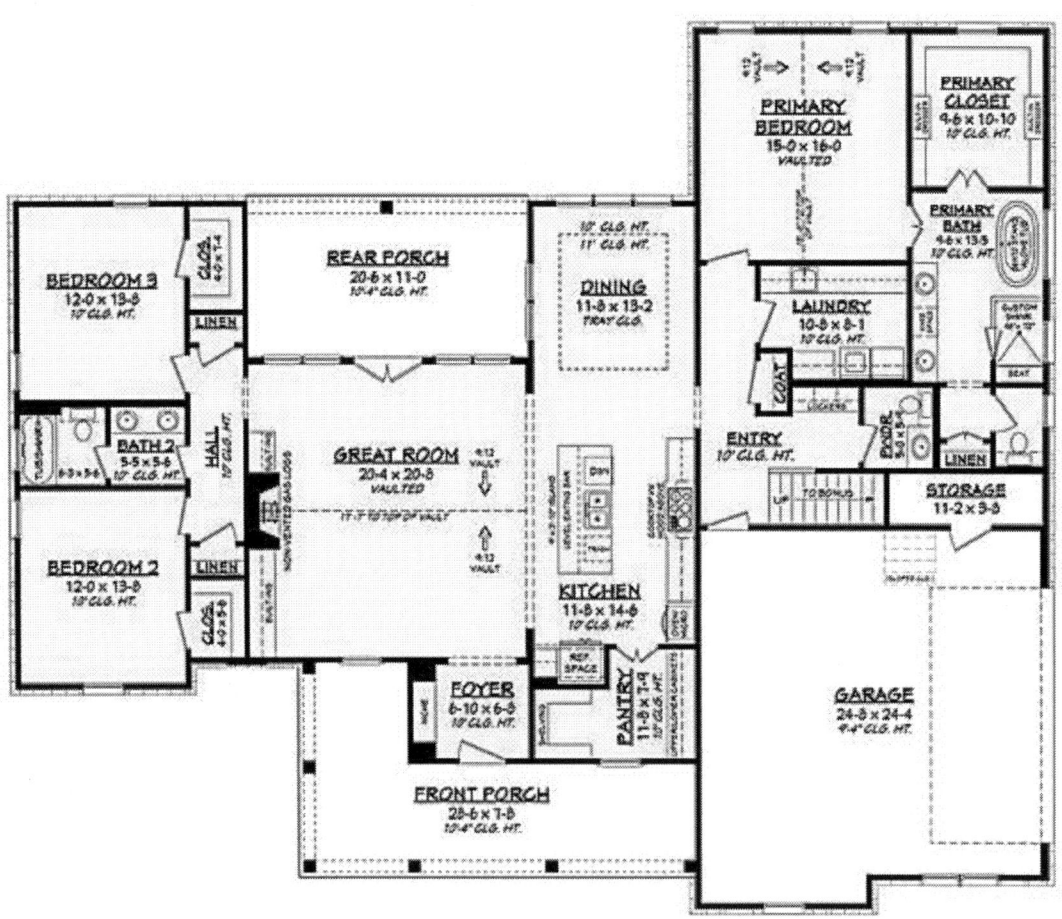

A vacation-type barndominium floor will focus on relaxation and provide facilities for recreation. The living rooms can be open and seamlessly integrated with decks outside, allowing a large space for social interactions and views to enjoy.

Homeowners could include in a home features of luxury that would allow it to become a transitional living style between inside and out by adding large kitchens, outside grilling areas, and large porches. Some of them would even include lofts or guest suites, making the area ideal for numerous families or large groups. Large windows and high ceilings in vacation homes can make travelers feel they are part of the landscape; this will create an easily accessible vantage for them to unwind and enjoy the scenery, whether it be a mountainous retreat, a lakeside cabin, or a timberland resort.

Cost-effectiveness and maintenance

Cost-effectiveness is one of the major characteristics that make barndominiums increasingly popular as permanent residences and for holidays. It could be that traditionally, residences are expensive to build and maintain, whereas a barndominium-true, especially of steel-framed construction comes with a relatively cheap price tag.

1. **Long-term home ownership**

Barndominiums stand out as an attractive option for investment in long-term home ownership, with lower construction costs and lessened maintenance when compared to other types of houses. Insurance premiums and maintenance costs become noticeably lower due to the resistance of the steel structure to most environmental factors like termites, rot, and fire.

Besides, many of the floor plans that have been designed for barndominiums are energy-efficient. These floor plans are generally strongly insulated with huge windows to help maximize the amount of natural light entering the abode, consequently minimizing the amount of energy consumed over time.

Solar panels, water-efficient systems, and other green technologies easily can be included in the design of a home for the homeowner who places much importance on sustainability.

2. **Advantages of a Vacation Home**

The aspect of low maintenance in a barndominium becomes more important when it is used as a vacation retreat, rather than a permanent residence. The owners can just sit back, enjoy their time out, and relax without much regard to its ongoing maintenance or repair.

The barndominiums are a great rental option for those who rent their vacation property when it's not in use because of the durable materials and how easy they are to keep up. Vacation renters who come through often look for lodging that is unique, yet practical, and the barndominium can provide just that with the rustic appeal and contemporary conveniences.

Family Size and Space: How much room will you need?

Perhaps the most important thing to consider building a house is the size of the living area, especially where the number of occupants in the house are much. It becomes even more crucial when it comes to the planning of a barndominium, especially one that offers a unique blend of modern functionality and natural charm.

From their open floor plans to their energy efficiency and cost-effectiveness, barndominiums mostly converted from barns into living spaces have gained tremendous attraction over the years. Like any other house, however, which room you will need depends on several factors, the size of your family being one of the most important ones.

This section looks at ways that family size can impact the amount of space needed for barndominium floor design and how one can better plan accordingly to yield comfort and functionality.

Understanding Barndominium Layouts

When it comes to barndominium floor plans, anything from small, simple designs to spacious layouts that contain several bedrooms and bathrooms and huge living areas is fair game. Most of the floor plans in these kinds of buildings are open, which provides greater flexibility in the use of space.

They are ideal for families because of flexibility; they can accommodate changes in needs over time. An open floor plan can make even the small houses seem much larger, but you need to have a good idea of how much square footage you'll need depending on family members.

Determining Your Family's Needs

To determine how much space your barndominium will require, you must first have an understanding of what the needs of your family are. Probably the most obvious determinant for the number of bedrooms, baths, and common living areas you will need is the number of your family members.

1. **Couples and families with a limited number of children**

Sometimes, for couples or families with only one or two kids, a smaller floor plan will work. With a barndominium containing two to three bedrooms and one or two bathrooms, most can afford a good number of living spaces.

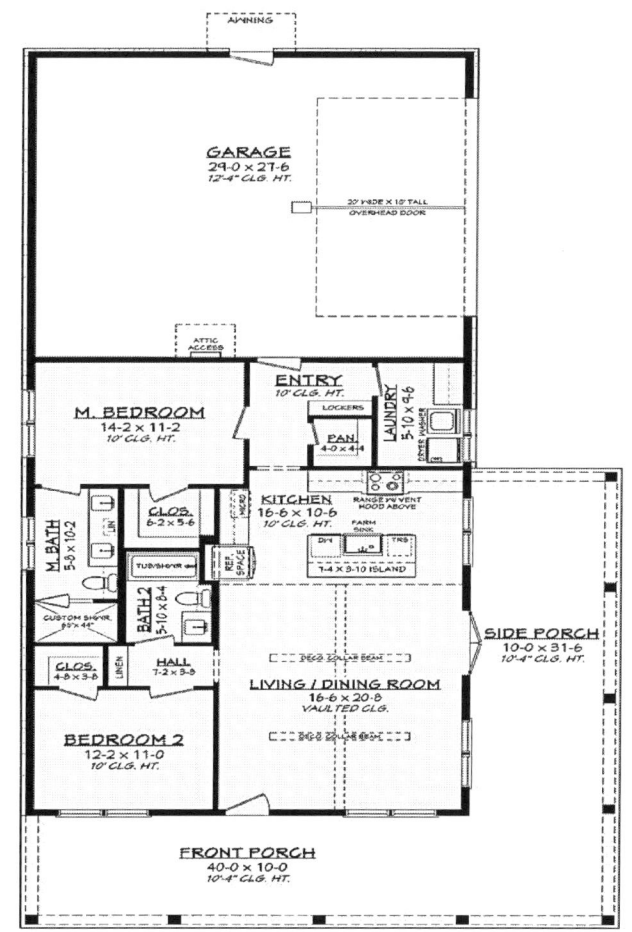

AWNING

GARAGE
29-0 × 27-6
12'-4" CLG. HT.

20' WIDE × 10' TALL
OVERHEAD DOOR

ATTIC ACCESS

ENTRY
10' CLG. HT.

LOCKERS

LAUNDRY
5-10 × 9-6

DRYER WASHER

M. BEDROOM
14-2 × 11-2
10' CLG. HT.

PAN.
4-0 × 4-4

CLOS.
6-2 × 5-6

KITCHEN
16-6 × 10-6
10' CLG. HT.

RANGE W/ VENT
HOOD ABOVE

FARM SINK

M. BATH
5-8 × 10-2

TUB/SHWR

REF. SPACE

DW

TRS

7-4 × 3-10 ISLAND

CUSTOM SHWR.
9'3"× 4'4"

BATH 2
5-10 × 8-4

SIDE PORCH
10-0 × 31-6
10'-4" CLG. HT.

CLOS.
4-8 × 3-8

LINEN

HALL
7-2 × 3-8

LIVING / DINING ROOM
16-6 × 20-8
VAULTED CLG.

BEDROOM 2
12-2 × 11-0
10' CLG. HT.

FRONT PORCH
40-0 × 10-0
10'-4" CLG. HT.

Smaller homes feature open floor plans that allow shared living spaces to flow seamlessly from one room to the next, thereby ensuring that all family members always feel connected to each other. With a family of this size, a floor plan ranging between 1,200 and 1,800 square feet provides the necessary comfort.

2. Medium-Sized Family Units

Families that are considered to fall into the medium category, either three or four children, require additional space for privacy and comfort. The need for more bedrooms and bathrooms, perhaps even an additional living room or family room becomes a necessity when trying to allow room for all family members to stretch out.

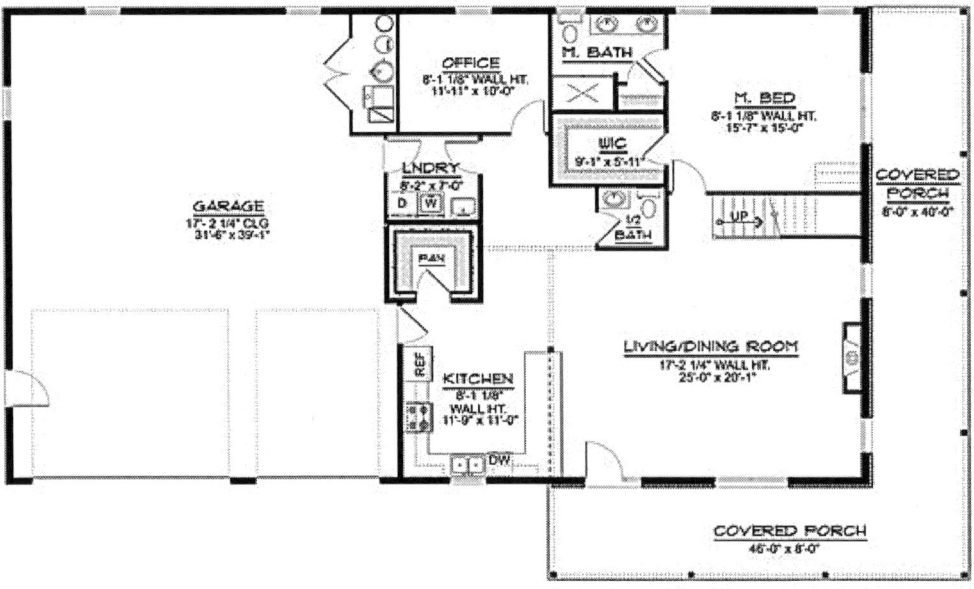

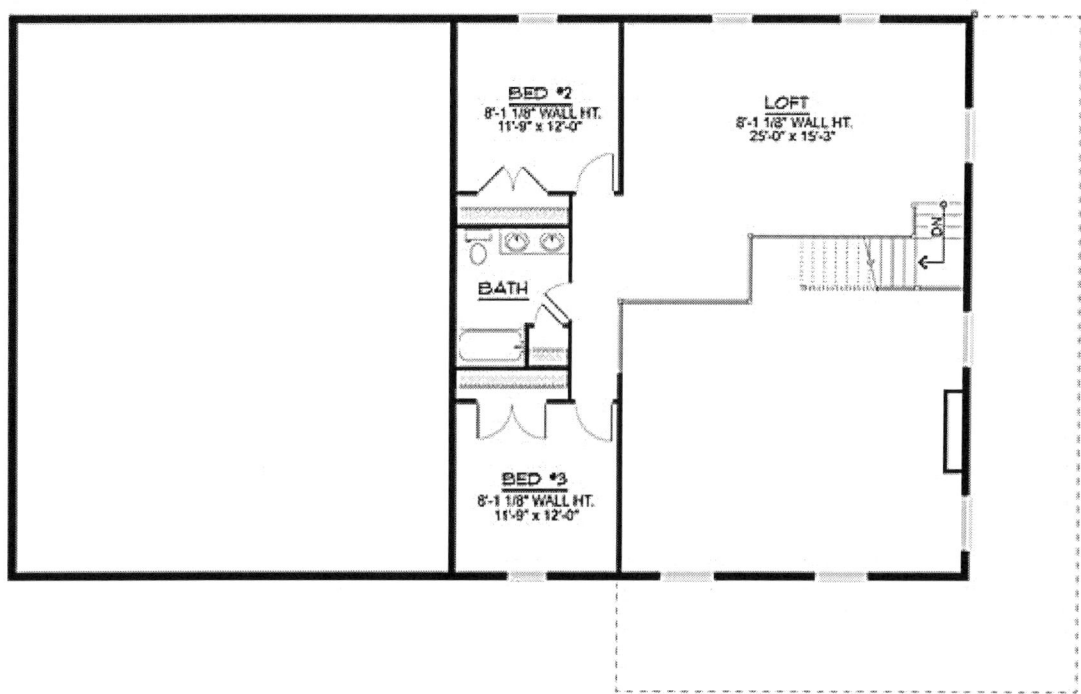

It is also recommended that, in addition to the two to three baths, the floor plan should contain at least three or four bedrooms. Furthermore, larger families with a pattern of preparing and taking their meals together might need larger kitchen and dining rooms.

It's not uncommon that a medium-sized family can be accommodated with a barndominium floor plan between 2,000 and 2,500 square feet.

3. Multiple generation or large families

Large families of five or more children or those who live with extended family members-grandparents-need a design that can allow for multiple private places to be accommodated and still maintain some sense of community.

More often than not, the preferred property options will be larger barndominiums than those at least 3,000 square feet or larger in size-featuring four or more bedrooms and several bathrooms, complete with multiple living areas.

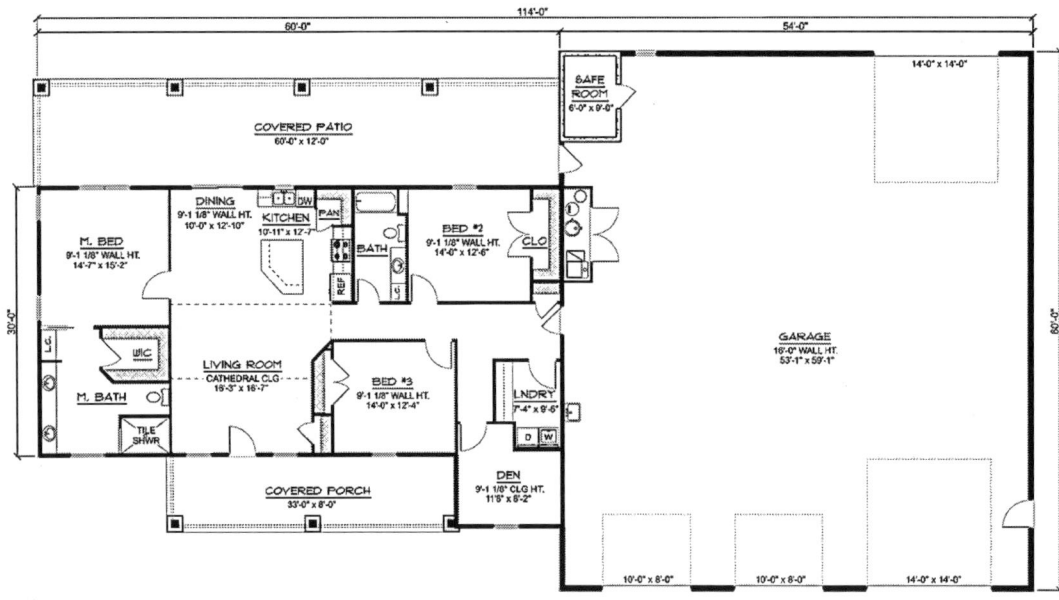

A home designed for multigenerational living can mean including a separate guest suite or apartment for in-laws so that everyone is under one roof, but there is some privacy for all.

Space allocation: key decisions for every room

When you're drafting your barndominium, you need to have a mindset for the spaces and how those will be put to use to best suit the needs of your family. Here's how some of these elements would need to be considered:

1. Bedrooms

A bedroom is perhaps one of the most evident places where the number of spaces required is directly proportional to the size of the family. Since people want privacy, it's almost a given that every person or every couple should have a room of their own, and the more family members you have, the more bedrooms you should have.

Apart from the number of rooms, you also have to take into consideration the size of the bedrooms. The master suite, of course, must be large enough to accommodate a king-size bed, clothes, and maybe even a sitting area. For instance, the bedrooms to be used by our children can be smaller. Personal touches like walk-in closets or en-suite bathrooms add to the livability of the house.

2. The Bathrooms

Another important factor to consider is the number of bathrooms within the dwelling. A less full house can easily get by with one or two bathrooms. In contrast, the demand for added bathrooms would increase proportionately with family size. Therefore, there should be one full bathroom for every two bedrooms, plus an additional half bathroom for visitors. This is generally an excellent rule of thumb.

If you are into a big family, then you have to consider having a master bedroom with an attached en-suite bathroom as you design your house to enhance the level of privacy and comfort.

3. Living Rooms

The living room of a house is the hub of the house because it is the room where the members of the family get together to share time. A single living, dining, and kitchen room that is open-concept may suffice for the smaller family.

On the contrary, additional living areas are an absolute necessity for a family of more members. This feature may include a second family room, a game room, or even a separate home office which can double as a study place for kids. The designs of barndominiums are characterized by open-concept living spaces; as such, the main living room should be designed to seat all the living members of the house.

4. Food preparation and dining areas

For those households who have enjoyed preparing meals and dining as a family, the kitchen and dining areas should be large enough to accommodate several people at one time over meals. It is okay to have a galley kitchen for families in smaller houses, but for those who have medium and large-sized homes, a more spacious kitchen with a huge island, plenty of counter space, and additional seating can help.

Due to this, the dining room should be large enough to accommodate all of them for sitting and making meal time an entertaining time for the whole family.

5. Storage and Usage Facilities

In residential design, decent storage is always neglected, although it is one of the most important features required in households where children are present. The larger the household size, the larger the need for storage space.

A well-built barndominium should have a closet in each bedroom additional storage closets scattered around the house, and even a pantry or utility room set aside for that purpose. Consider adding a mudroom into your home in which to store shoes, jackets, and other items that are used regularly.

How to Future-Proof Your Barndominium

Regarding the floor plan design of your barndominium, you shouldn't just look at your current needs, but your future needs. Your use of space may change as your family expands and develops. For instance, you may find yourself needing to accommodate young children having their room, and even your elderly parents coming to live with you later in the future. An open floor plan presents the ability for future adjustments since one can add walls to divide it into rooms or repurpose a room in response to changing or evolving needs.

That being said, you'll still need to consider how your lifestyle may shift. You'll need a home office or study in case you work from home or homeschool your kids. In case you are the type who enjoys entertaining friends and family, having enough entertaining rooms in the barndominium will make it a more functional space.

Budget: How your floor plan affects the overall cost

A floor plan is one of the most integral parts in estimating how much money you are going to have to spend on the construction of a barndominium. The layout and design of your home will have a direct relation to the cost that is spent on constructing it. While planning, you need to understand just how the floor plan options available will affect your budget.

Some of the major components of the cost of construction for a barndominium floor plan include size, complexity, material choices, and functional spaces. In this section, we will break down important elements that influence the overall cost of building.

1. **Bigger size and square footage increases the overall cost**

Of the various ways that your floor plan can impact the total cost, probably the most obvious approach is in size. As the square footage of your barndominium goes up, so too does the total amount of materials and labor to construct it, thus going up in overall cost.

As much as a big size of home may give you space for living and storage, it requires a bigger investment to enable you to build it successfully. The overall square feet of your barndominium comprises not only living areas but also porches, lofts, and storage facilities among others.

It's important to realize that this number is not a fixed cost per square foot; rather, it's going to vary by a great degree with the certain materials and features you may choose. For the rough estimate, this might go from $100 to $150 per square foot for a basic and practical floor plan to a bespoke luxury design incorporating high-end finishes and features at up to $200 or more per square foot.

A helpful hint: When deciding on the size of your floor plan, keep in mind what your needs are. If you are willing to stay within a limited budget, you will want to decide upon a more compact, efficient layout that uses space with as little square footage as possible that is not necessary.

2. **The Complexity of the Floor Plan: Keeping Things Simple Will Help You Save Money.**

Another very crucial component of the design in determining the cost is the level of difficulty it presents. A simple floor design that takes a square or rectangular shape is cheaper since it is easier and faster to build. The framing, roofing, and the building of the foundation require fewer materials. Besides, the number of sophisticated cuts or angles probably needed for a more complex design decreases, as well as the energy consumption.

Conversely, it will be more expensive if the floor design is very complicated, with many wings, roofs with different pitches, or including some special features. In addition, these components require not only more specialized labor and materials but also more time to complete, adding to the costs involved with labor.

While crafting a minimal budget, it is best to go with designs that are simpler and have fewer detailed structural components. Barndominiums are indeed known to be versatile even when offering a rustic-modern look with clean lines and open spaces since simplicity in design does not equate to sacrificing style.

3. The option to choose between functional and aesthetic materials.

The most critical factor affecting the price of your barndominium will be the material you choose. Designs of the floor using less expensive materials for framing, siding, and roofing, such as metals, will be cheaper than those designs relying on more traditional materials like brick or wood. Since barndominiums are often constructed out of steel, they tend to come cheaper compared to conventional dwellings. This is particularly the case when considering a rural or agricultural location where access to steel and other materials is readily available.

The prices will increase once your floor plan involves hardwood flooring, granite worktops, custom cabinetry, or designer windows. When you are designing your floor plan, it is about finding that happy middle ground between the functionality of the space and aesthetic appeal. Given that metal is durable, low-maintenance, and fire-resistant, it is a material that one could use cheaply; however, you might prefer to include more luxurious materials in key parts of the home, such as the kitchen and bathrooms, or living areas, whether out of preference or budget.

It is suggested that for most of the construction, you choose material with a long life and low maintenance if you are concerned about your budget. You may want to save some extra bucks to buy luxurious features or finishes in the main parts of the house, which will enhance the whole look and appeal of the home.

4. Framing and Finishing Costs: Open Concept Spaces and Partitioned Spaces.

The open-concept floor plan that minimizes interior walls used within the home is usually one of the most appreciated attributes of barndominiums. Where there are fewer walls, there are fewer materials involved and also fewer laborers, thus the overall building costs can be considerably reduced. Barndominium development: most people enjoy open floor designs for the sole reason that they make a space feel so much bigger and give a sense of continuity or flow.

On the other hand, compartmentalized areas are great for families because of privacy concerns or functional use. This is going to involve extra framing, drywall, and finishing touches adding more costs to the overall price of the project. As a general rule, if there are more 'closed' spaces in your floor plan, such as separate dining areas, offices, or even spare bedrooms, the costs keep going up. Bathrooms can be another area of added costs, given that they require plumbing coupled with specialist finishes such as tile and waterproofing.

Tip: If you like the idea of open spaces but need your privacy, then you might want to look for something called a hybrid floor plan. A hybrid floor plan combines an open main area with private, enclosed bedrooms or office areas. You can have a balance between comfort and cost with this approach.

5. **Energy Efficiency and HVAC Layout of the Floor Plan.**

Remember to put some extra consideration into your floor plan selection for energy efficiency when designing your barndominium. Large expansive areas are harder to keep a consistent temperature in, leading to high heating and cooling costs. With barndominiums featuring vaulted ceilings or wide windows, the views can be spectacular-looking, but if not well-insulated or fitted with energy-efficient technologies, they can also lead to higher utility bills.

Moreover, the floor plan of your building directly relates to the setup of your HVAC system configuration. A more complicated floor plan with a higher number of rooms, corners, and fluctuating ceiling heights may call for a similarly complicated HVAC system. This adds more costs not only to the installation but also to the operation in the long term.

With the above said, some of the popular HVAC options for barndominums include;

- **Mini-Split systems**

- **Traditional central HVAC systems.**

- **Heat pumps**

Advice: If you're trying to save bucks on using money for heating and cooling, go for an energy-efficient open floor design with insulation and window placement. That is why you will not only save your dollars in the initial construction but also energy bills afterward.

6. Practical areas: garages, lofts, and other facilities.

Many of the floor plans for the barndominium condos come with functional spaces such as garages, lofts, or workshops. While such added features might considerably enhance your property in its utilitarian value, they will certainly raise the bottom-line price. For example, adding a loft means more material use for flooring, framing, and structural support. Equally, large garages or workshops will need additional electrical wiring, plumbing, or specialty flooring added to the increased square footage of the room.

The hint here is to think about which functional spaces are essentially needed for your lifestyle and then incorporate those spaces into the floor design in the right manner. It might be cheaper to add these later rather than include them in the construct itself as part of an expansion project.

Building Codes and Zoning: What permits and regulations must you adhere to?

Since barndominiums are a hybrid, and also because they very often fall under the category of rural buildings, constructing a barndominium requires a unique blend of regulations and permissions that come under both residential and agricultural construction standards. A decent understanding of zoning and construction standards is, therefore, fully necessary to ensure compliance and prevent delays, fines, or even legal complications.

1. Acquiring Your Knowledge of the Building Codes for Barndominiums

Building codes are a set of rules meant to ensure that a building remains safe in all instances as well as structurally sound and functional. When erecting a barndominium, such a set of building regulations may differ since they are based upon the locality where the building is to be constructed, the purpose to be served, and the materials used. While individual codes do indeed differ from state to state, or even city to city, there are several general areas of concern that most jurisdictions have in common.

a. IBC and IRC stand for International Building Code and International Residential Code, respectively.

Most building regulations of municipalities in the US are formulated from two important sources: the International Building Code, IBC, and the International Residential Code, IRC. While the latter prescribes residential homes of one or two families, structures applied to the IBC are of either commercial or mixed-use types.

Because barndominiums often incorporate both residential and non-residential uses- such as barns or workshops-both codes will likely apply to certain features of each. As an example, the residential portion of the barndominium would have to meet the requirements within the IRC for insulation, electricity, plumbing, and fire safety. The agricultural or workshop sections may fall under other codes, as determined by the IBC or other specific construction codes for agricultural buildings.

b. Requirements Related to the Structure.

Whatever the code followed, the basic structural safety parameters are sure to be met by the barndominiums. To determine that the building can support any local conditions relating to snow load, wind speeds, or seismic activity, you will be required to submit specific blueprints showing the design and structural calculations involved to the local building authority.

Additional standards may be applied to accommodate hurricanes or earthquakes, but again, it would depend on the area. For example, houses located in coastal areas may be required to be built such that they can withstand high wind conditions. For areas prone to earthquakes, you could be forced to make use of special foundations and bracing.

c. Electrical, plumbing, and mechanical systems.

Similar to any other residential construction, the barndominiums are required to adhere to the local dictates regarding electrical wiring, plumbing systems, and mechanical installations, which also include heating, ventilation, and air conditioning, better known as HVAC.

More often than not, such systems are required to be installed by licensed contractors and to pass while the building is still under construction. Following set rules, plumbing systems will be interconnected properly, electrical outlets suitably spaced, and heating and cooling systems scaled and built properly.

d. Maintaining Fire Safety.

Barndominiums are hybrid constructions so fire safety is something very crucial to consider. According to the size and construction of the living quarters, residential areas may or may not require smoke alarms, fire extinguishers, and sprinkler systems.

In some odd cases, sprinkler systems may also be required. To the degree that the barndominium has large open areas, such as barns or workshops, the need for added fire safety may be required to make sure areas of this sort are not a danger to the dwelling parts of the building. Examples include the use of fire-resistant materials, fire doors, and appropriate storage of flammable products.

2. Zoning Regulations Applicable to Barndominiums.

Zoning ordinances are laws that regulate land use. These are restrictions as to where specific structures should be constructed and how the property should be used within the constraints. It is the zoning restrictions that determine whether or not the barndominium can be constructed on a specific lot, its size, location, and use. In most cases, these are regulated by local governments, and the actual regulations may be very different from one jurisdiction to another.

a. Three categories of zoning: residential, agricultural, and mixed-use.

One of the first steps in planning your barndominium is to determine your property's zoning designation. A lot of the time, the land in rural areas is zoned for farm uses, which in turn can make it easier for permission to be granted for a structure to be built on the land, like a barndominium. Conversely, if the land is strictly zoned for residential use, any facility building, such as a farmhouse or workshop with areas not dedicated to being residential, could well be hampered. On the other hand, if the land were to be zoned for agricultural purposes, there would be limitations to building any dwelling or residence that is intended for full-time use.

There are those circumstances where a mixed-use designation would be required, or you may be required to apply for a zoning variance to allow both residential and non-residential uses of the property. The whole process can be extremely time-consuming and may involve public hearings or local planning department reviews if the subject site is not currently zoned suitably.

b. Setbacks, Lot Coverage, and Height Restrictions.

Setbacks, the minimum distances a building must be from property borders, roads, and other structures, are other standard requirements within most zoning regulations. Other regulations of zoning ordinances include setbacks. Setback regulations intend to position buildings correctly for fire safety, privacy, and environmental reasons. With these regulations, you will need to adjust where your barndominium sits on the land. This may be depending on the size of your lot and the regulations that are within your area.

Mainly, a large portion of zoning laws will also include limitations regarding how much of the land can be covered with buildings. This is commonly referred to as lot coverage. If you have plans to build a big barndominium in the same lot where you will have additional barns, garages, or workshops, this is what you should consider seriously. If you are planning on having any lofted rooms or tall barn buildings in your design, you need to know that local height restrictions can affect your design, too.

c. Permitted Uses and Occupancy.

Also, zoning restrictions will define what the building will be used for, which may or may not restrict whether you can live in the barndominium full time or it can be allowed only as a second residence or summer home. Municipal legislation in some agricultural-based locations would restrict home occupancy or would put constraints on how frequently the structure could be used for dwelling purposes. These may likely affect your plans if you want the barndominium to be your main house.

3. Building a barndominium requires securing permits

Once you know your site is appropriately zoned and that your design is acceptable under the building codes, you'll need to obtain many permits before you can begin actual construction. These are;

- **Building permits:** Obtaining the construction permits is the sure way to make sure that your structure is up to code on all the building codes. You will be asked to provide complete floor plans with structural drawings and engineering calculations.
- **Electrical and Plumbing Permits:** These permits ensure that all plumbing and electrical systems are installed according to the local safety code and that they are checked at various stages in the construction process.
- **Zoning Permits:** Where you are may also dictate that a different type of permit, termed a zoning permit, is necessary solely to ensure that your project is compliant with the local land-use restrictions.

- **Permissions with environmental agencies:** In cases where your barndominium will affect wetland areas, wildlife habitats, or rivers, there are permissions with environmental agencies that you may need at either the local or federal levels.

How to choose the correct land size and location for your barndominium

Choosing the right land size and location is one of the biggest decisions you will have to make in the planning process. A barndominium is a unique, versatile abode that fuses interior space with outside features suggestive of a barn, therefore offering a rustic and spacious ambiance. Whether your heart's desire is a modern open-concept barndominium or one in a more traditional design, the right plot of land makes all the difference between your house serving to complement your lifestyle and further value your property in the long term.

Below is a detailed guide to assist you in selecting the right land size and location for your barndominium while giving you key considerations about floor designs and future requirements.

1. Knowing Your Barndominium's Needs

Before starting the land search process, a buyer needs to have a clear vision of what type of barndominium they will purchase. The term "barndominium" can refer to everything from a simple, one-story dwelling to an elaborate, multi-story home with garages, workshops, and outdoor living areas.

As you plan your barndominium, here are some key questions to consider:

- How many square feet do you require to live?
- Do you need a garage, barn, workshop, etc. space?
- Would you have any outdoor activities such as farming, gardening, or leisure spaces?
- Would you want a one-story or prefer a two-story design?

By answering these types of questions, you'll be able to determine what square footage best suits your needs in a home, and also what amount of land will be required to support that home. You'll want more land if you plan on building a large workshop or garage area, for instance than someone who just wants to build a smaller living space.

2. Choosing the Right Size of Land

Several factors will determine how big your piece of land will be, including the floor plan of your barndominium, the laws that determine zoning, and your personal preference.

Things to Note about Land Size:

- **Zoning Laws, and Building Codes:** First, find out about any local zoning ordinances that dictate where you can build. Some areas might have minimum acreage requirements, restrictions as to the types of buildings you can build, or even regulations on how far your house must be set back from the boundary of the land. Your knowledge of these will save you from some legal headaches in the future.

- **Lot Size for the Floor design**: If your barndominium is going to be bigger than a typical residential home, then you will want to ensure that the land is big enough that fit the floor design inside in a comfortable manner. For example, if your floor plan calls for a home that is 3,000 square feet in size and also contains a garage and a workshop, one-acre plots of land may feel too little. Larger houses with other accessory buildings, such as barns, stores, or stables, may be best suited for at least two to five acres of land to accommodate what needs accommodation.

- **Future expansions**: You need to think about the future of your business. Even if you are not planning to build a barn shortly or add an extension to your house, buying enough acres will be wise for whenever any future expansion possibilities crop up. It becomes very important whether you plan to use the land for agricultural needs, gardening, or other recreational activities like horseback riding.

- **Privacy**: There are various, possibly hundreds of, reasons why people choose to live in a barndominium. Probably one of the critical ones will be the privacy it allows them to have. Put a price on privacy, and you'll find owning a bigger plot of land allows sitting your home in such a location that would be remote from your neighbors and major roadways. A lot that is at least three to five acres in size can provide enough buffer space, which guarantees peace.

- **Topography and Land Use**: The size of the land you will need is further affected by the actual physical characteristics of the land itself. Does the land have hills and valleys on it, or does the land have a flat surface? You may want more acreage because you need the land to have steep inclines or the terrain to be rocky, for example, to support using the buildable section of the property for your barndominium floor design. In addition, think about how you will use the land. For example, if you intend to raise cattle or crops, then you will want more acreage than a person who will only need a yard and garden area.

3. **Location, Location, and More Location**

Equally as important as the amount of land you purchase, is where your barndominium will be located. The location of the land affects your lifestyle, how much money you will put into construction, and even the long-term worth of your investment.

Things to Consider While Choosing the Right Location

- **Convenience to Amenities:** You need to know how far you are willing to be from essential amenities like grocery stores, hospitals, schools, and jobs. Maybe the rural site offers peace at the mercy of being farther away from such necessities. The more rural it is, often the cheaper land with fewer building restrictions you can find if you do not mind a long drive to commute.

- **Amenities and infrastructure:** The land should be connected to services like electricity, water, sewage, and the internet. This could be pretty expensive if the land is more remote. It's also a good idea to check road access, which simply means that the subject land is connected to a public road or an easement for driveway construction.

- **Environmental Issues:** You will want to research environmental concerns like flood zones, soil quality, and drainage. You won't want to invest in the property only to find that it tends to flood; this can be very harsh on the structure and foundation of your barndominium. In addition, poor-quality soil may affect your ability to have a garden or landscape on certain land.

- **Resale Value**: The location of your barndominium will also determine the price at which it is being resold later on. You stand a chance whereby, upon building in an area where other homes of a similar design are also being put up, the value of your property might improve with time. Conversely, one that is utterly disconnected from the rest of the world may raise some obstacles in case you try to sell your property later on.

- **Community and Lifestyle**: In case you are thinking of building a barndominium, then there is a need for you to visit the community where you will live. How do you like the surroundings? How warm-hearted are the people around? Is the area growing, or does it remain the same? By knowing what goes on in the local culture and way of life, you may be able to determine if you will be happy in that area long-term.

4. Working with a Knowledgeable Person

Determining the size of land and place for your barndominium may be a little tricky, especially if you are unfamiliar with how this works. It will be a great advantage to engage in services with a real estate agent who has had past experiences in dealing with barndominiums and rural properties.

You'll get to handle zoning restrictions through them, find the size of the property you need, and find a location that works for you. Additionally, the architects or builders who have experienced barndominium designs will also be able to advise you about how much land you will need, depending on the one-floor plan that you may have chosen and your future ambitions.

CHAPTER THREE

Common Features in Barndominium Floor Plans

Many of the floor plans for barndominiums include open-concept designs that incorporate the living sections of your home, which will include a kitchen, dining area, and living room in one huge room shared by all its residents. Such a plan maximizes available space and installs a sense of community. It is commonly associated with high ceilings and larger windows, further giving that airy feeling inside the interior. Large, multi-use rooms can also be found in barndominiums from time to time, like huge garages or workshops created to answer the needs of their homeowners for space or facilities that can accommodate hobbies.

Other common characteristics to be shared in all barndominium floor plans are flexibility and personalization. Many of the floor plans are designed in a manner that caters to the needs of a homeowner with flexible space to be used for other purposes, such as additional bedrooms, home offices, or entertainment. These metal houses are also long-lasting structures of the dwellings and would make expansion or future adjustment of the dwellings very easy. Such homes were meant to integrate the rustic charm of the past along with the conveniences of the present, and they often include huge outdoor living areas like wraparound porches along with energy-efficient technologies.

Open Floor Layout: Open and Versatile.

Amongst the current trends for house design, the open floor plan is still at the top of the list. It provides owners with a spacious and flexible area. Probably the most impressive feature of barndominium homes is the open floor plan. An open floor plan allows for more flexibility in design and construction. The open concept creates flow, makes the space appear larger, and creates a very social and connected way of life.

In this section, we go into detail about the many pros and features that come with open floor plans, specifically in regards to barndominium floor plans, and how this design method can marry practicality with aesthetics.

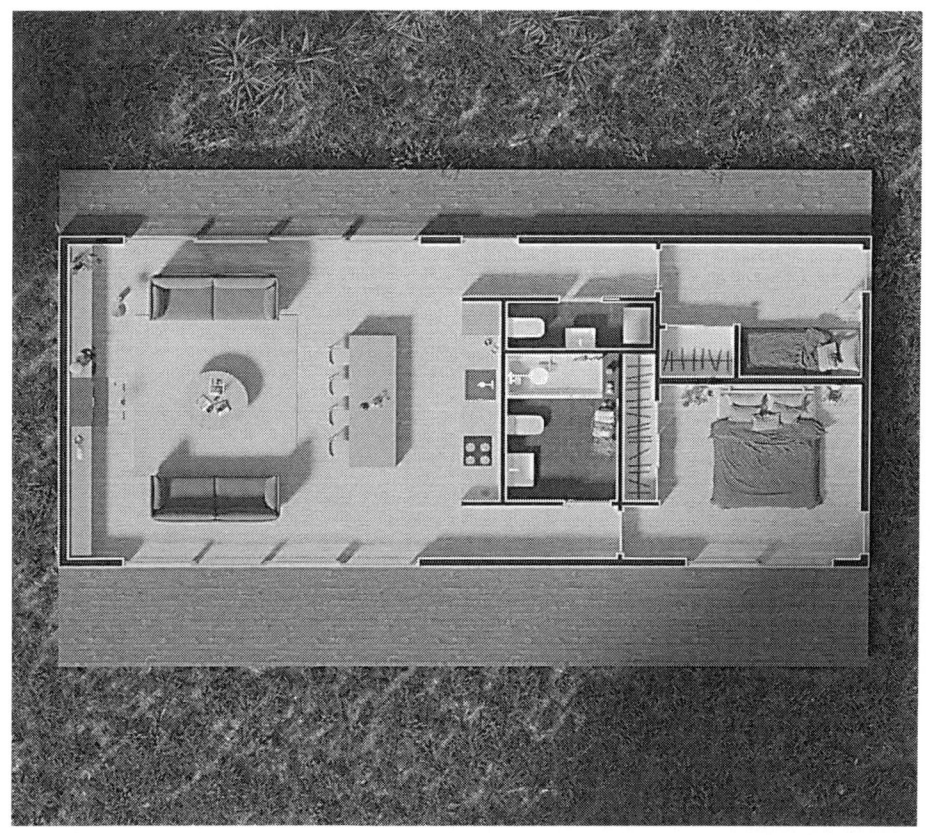

So, what is an open floor plan?

By open floor design, it means that the architecture of the home is one in which different common areas of the living room, kitchen, and dining area merge into one huge area that is open and continuous, bereft of any separating walls. An open layout provides a living atmosphere with no obstructions in the flow, compared to compartmentalized rooms for every purpose. This architectural method applies to most modern houses, aiming to grant more space and let more natural light in.

Barndominiums are the perfect target for an open floor plan since all of them boast huge floor plans and expandable structures. At the very beginning, barns were meant to be huge, open areas that could be used for storing crops, cattle, or machinery. With the open concept, this big space has the potential to be converted into a modern living area that undergoes conversion into homes.

Advantages of an open floor plan in the barndominium include:

- **Improved Size and Flow**

The most important advantage of the open floor plan adds up to how spacious it feels. The removal of walls and partitions creates the impression of a larger, more open whole house, though the square footage is quite small. This design realizes most of the naturally enormous internal space present in a barn home creating the impression of being spacious and airy. The kitchen, dining, and living room spaces are all connected, allowing residents to cross over easily. This quality makes the space ideal for families or those who enjoy parties.

- **More Natural Light**

The open floor plan allows sunlight to light up a room easily, which is excellent in a barndominium that has huge windows or sliding glass doors. Light can flow from one end of the house completely to the other, illuminating every area of the house, since there are no barriers to impede the light. This makes the setting more appealing; at the same time, there is less need for artificial light during the day, which again is a plus for energy efficiency.

- **Flexibility in Interior Design**

The presence of an open floor plan provides great flexibility in terms of interior design. Since there are no walls, the homeowner has more flexibility in how they would like to set up furniture and other features in congruence with their lifestyle or preference.

Most of the time, the main living space in a barndominium is one huge open area that can translate into an infinity of innovative design implementation opportunities. With an open plan, you get to try all sorts of different styles and layouts to achieve the homey rustic or sleek and contemporary look. Furniture, carpets, and lighting could be used to create different areas or zones within an open space without actual physical dividers, all to set apart the various sections within the open space.

- **Increased Social Interaction**

An open floor plan allows the occupants to be connected across the house as they interact with one another for social reasons. In a typical house design where the house has different rooms, the kitchen, living and dining are usually separated from each other.

It limits the members of the family or visitors from communicating effectively with one another, especially when they are doing different things in different parts of the house. An open-plan barndominium presents the opportunity for an individual who is cooking in the kitchen to converse with other people in the living room or dining area. It is something people who love to plan activities or simply spend quality time with their loved ones would appreciate due to the seamless connectivity between the rooms.

- **Space Usage Optimization**

Barndominiums are indeed known for their flexibility and also their potential when it comes to space optimization. This is further enhanced by the open floor concept eliminating waste areas such as corridors and little rooms not serving their purpose. The flow between spaces ensures that every square foot serves a good purpose; this again makes the home more useful and efficient because it is effective.

It is a great option for those homeowners who are trying to downsize their home or trying to build a home that is more ecological and minimalist to have an open plan in a barndominium.

- **Customization Options**

One of the key features the barndominiums possess is the fact that they can be customized. Homeowners might also work with an architect or builder to design a floor plan that suits their specific needs and wants.

Since there are fewer structural restrictions on an open floor plan, there are also more options for the homeowner to construct a space that better aligns with their vision for the area. It is a venue that can be remodeled using lofted rooms, customized cabinetry, and even a central kitchen island at the owner's discretion for creativity.

- **Accessibility Increased**

Also, open floors are quite functional regarding accessibility. People with certain mobility problems or wheelchair users will more easily manage the area due to fewer barriers and obstacles in the general view. The wide-open areas representative of a barndominium can be constructed to be barrier-free; thus, creating an atmosphere friendly for all of the residents.

Open Floor Barndominium Designs

- **Centralized Kitchen**

A central kitchen is perhaps among the most common architectural features known and shared by many barndominiums that have open floor designs. Placing the kitchen at the core of the house makes it one focal point that usually brings family and friends together in socializing or bonding. A large kitchen island can add more seating and counter space to the room but also help visually to be a boundary between the kitchen and the rest of the living rooms of the home.

- **Lofty Spaces**

A lofted room can be added to a barndominium that has high ceilings to create additional space to live or store without interrupting the open feel of the structure. The lofts can serve as bedrooms, home offices, or entertainment areas and let the rest of the home stay open and airy.

- **Separated Areas Furnished with Furniture**

Even if there are no boundaries, an open floor plan can also define different zones with the use of accessories and furniture. Distinctive functional zones inside the open space can be marked with the use of sectional sofas, area rugs, and strategically positioned lighting.

Flexible living spaces that can easily adapt to different uses.

The 21st century brought with it changes in the way we work and live: increasing numbers work from home, family units are changing in size and composition, and personal pursuits become ever more varied. In this way, there is an emerging demand for homes which easily adapt to a variety of uses. Barndominiums lend themselves well to this due to the general original open floor plan that can easily be changed around to suit each person's specific needs.

It can be as simple as a barndominium going from being a normal family home into devoting one room to a home office, studio, or even small business. It could easily be that the same room is used as a yoga studio today and a workshop tomorrow, depending on how the homeowner decides to structure the open space. Variety is one of many reasons barndominiums are becoming increasingly popular.

Flexible Barndominium Floor Plan Design

When designing a flexible floor plan for the living space of a barndominium, the options are virtually endless. There are, however, particular methods that can best make a space adaptable while remaining cohesive and functional. Some of those methods will be discussed here. In designing a living area to be flexible within a barndominium, the following can be considered:

- **Open Concept Layout**

By definition, a barndominium has an open-concept floor plan. One of the many reasons they are inherently flexible is that open spaces can easily be repurposed based on the intended usage, which will vary according to furniture configurations. For example, one large great room could function as a living room, dining area, or play/recreation room based on the furniture selected and how it is set up.

This setup removes the walls that might otherwise, under other circumstances, constrain the use of space. By keeping things open, you allow the room to be easily adapted to suit needs that change with time.

- **Mobile Partitions and Furniture**

Perhaps another major component of flexible barndominium floor plans is the inclusion of mobile partitions or sliding walls. These architectural features allow for the possibility of partitioning a large space into several smaller rooms when one feels this is necessary.

Sliding barn doors or mobile bookcases can be used, for example, to provide seclusion in a home office or guest bedroom and quickly move out when more open space becomes desirable.

Besides, furniture can help contribute to distinguishing the various areas that are encompassed within the barndominium. Pull-out couches, Murphy beds, and modular sofas are a few of the top examples of multi-functional furniture that can be changed to change the flow of space and even the functionality of the pieces of furniture. Doing so, allows the floor plan to be flexible and malleable since the space can easily convert between functions without having to conduct expensive renovations.

- **Zoning for Multi-Purpose Spaces**

One of the most efficient ways to ensure there is flexible living in a barndominium could be smart zoning. Zoning refers to an open floor plan, which is divided into specific "zones" that are allocated for certain purposes.

Having a kitchen that flows easily into a dining or living space is one example of how these zones can overlap regarding their functions. Some areas can be given over to more specific activities, such as a workshop or gym, but making sure mobility exists between these zones allows for a more dynamic lifestyle.

Zones in this general space can be softly delineated by utilizing multiple ceiling heights, special flooring, and specially designed lighting fixtures. This technology not only allows for the ability to set up visual boundaries without permanent barriers, but the utilization of the facility allows the homeowner to transition between activities with ease.

- **Loft Spaces for the Extra Flexibility.**

Many of the barndominiums include loft areas that add another layer of versatility to the floor concept. Lofts are great to make into an extra living area as a guest room, reading nook, or office because this area is suitable for any of those uses. Lofts use vertical space, so additional square feet do not have to be added to the footprint of the property.

This is an effective and economical method to add more capability to a flexible floor design. The loft could be converted depending on the dynamics of the family or the demands of lifestyle, from a children's play area to a quiet adult space, for example.

- **Integrating Workspaces**

Because the trend of remote work and freelancing professions is constantly rising within the workforce, the need for integrated workspaces in the house has been felt acutely. A barndominium can be designed with specific offices and studios in mind, even to the point that those spaces could be adapted to other uses if needed.

For instance, a workshop might serve as a garage or storage area, and an office could serve as a guest room if family members should come to visit. It is the hallmark of flexible living space to be changed into circumstances surrounding the occupant, and the barndominiums provide an ideal environment for that.

Lofts and Multi-Purpose Rooms: Use space to the fullest

Lofts are only a good option in maximizing the vertical space in a barndominium, especially considering that most of them have extremely high ceilings. Besides, one definition of loft is "an elevated area which may be used as a living space, office, or storage space." Lofts are quite versatile. It's a concept that's completely in tune with the open and airy aspect of the barndominiums but also provides extra square feet without extending the building's footprint.

Lofts in smaller barndominiums, where space is critical, are usually used either for sleeping quarters or guest rooms. This happens to be more accurate in tiny barndominiums. Having bedrooms in lofted spaces allows the owner to keep the main floor open and wide, making it perfect for large rooms that have multiple purposes such as living areas, kitchen, or living areas. The loft, with its heights, is an excellent choice to make a warm and secluded retreat even when the loft is not completely enclosed.

In the larger barndominiums, lofts can be used for any number of other purposes beyond that of a bedroom. They can serve as home offices, playrooms, or hobby rooms. The most interesting thing concerning a loft is perhaps the fact that it can easily be adapted and used in many other ways as time goes by. For example, a loft used initially as the children's playground can easily be turned into either a place to study or a home office when these children grow up. Lofts are among the trendiest barndominium designs for many reasons; versatility included.

Flexible Living Revolving Around Multi-Purpose Rooms

In building barndominiums, flexibility is always one of the main concerns; therefore, multi-purpose rooms will always be an important ingredient of this concept. While traditional houses are designed to have room purposes, barndominiums are open-concept in nature and can be changed as the day goes by and throughout the years. Since these multi-purpose rooms can be built to adapt to the needs of the homeowner, they are considered part of space optimization.

For example, a wide-open area on the main floor can function simultaneously as a living room, dining room, and home office when furniture and room dividers are intelligently placed in appropriate places. Sliding doors or barn doors can be installed to close off areas when privacy is required or to keep the feel of the room open and large if it is not. This is particularly helpful for families, where one room may serve as a playroom that then becomes a study or craft area as children grow older.

Besides that, multi-purpose rooms are suitable for holding events and meetings more frequently. In fact, having an open space allows you to accommodate guests easily by making changes to furnishings according to their needs. Their ability to adapt alone would turn multi-purpose rooms

47

into ideal spots not only for those who throw parties frequently but also for any kind of user for allows an infinite number of parallel activities in the same physical space.

Approaches to Design that Achieve Maximum Versatility

Lofts and multi-purpose rooms in a barndominium require very careful planning to make effective use of available space. Following is a list of things one may want to remember in designing the floor plan of a barndominium:

- **Use Available Vertical Space:** Since most barndominiums are designed with high ceilings, you should ensure that you use the available vertical space. Most barndominiums are ideally meant to have lofts. Think about what you will be doing in your loft and how much headroom will be required. While a sleeping loft may be acceptable with minimal headroom, you will want and need much more clearance in a home office or living area.
- **Incorporate Flexible Sorage:** Multi-purpose areas also call for flexible storage, because multi-purpose spaces need a minimum amount of clutter to make the space work. The inclusion of hidden storage, such as built-in cabinetry, under-bed storage, or closet organizers, allows for quick and easy transitions between uses. For instance, a living room with hidden storage can easily be converted into either a home office or an entertainment area without clutter from too much junk.
- **Plan for Natural Light:** With the very nature of a barndominium being to host large, open areas, it would go without saying that natural light plays an important part in the design of the building. This is possible with the installation of wide windows, skylights, and glass doors; it allows the inside to flood with light. It creates an illusion that this particular area is larger and much more inviting. In some instances, in an open, spacious floor plan, a loft can be well-lit and give the impression of being a different place altogether.
- **Create Zones with Furniture:** Zoning by furniture can help overcome one challenge or another of keeping a sense of order and usefulness in multi-purpose rooms. When correct zoning is absent, open floor patterns sometimes can create an impression of being overwhelming. Smartly using furniture may be able to define several sections within the available space. For instance, a sectional sofa might be employed to separate the living space from the dining area, while a large dining table might serve as a work surface during the day.
- **Install Sliding or Barn Doors:** Sliding and barn doors are very nice features for multi-purpose rooms; they provide room for flexibility without taking up too much space because they are easy to install. They could promptly become a private office or a guest room by rapidly converting open space, when necessary. Indeed, this is very functional, more so in the case of barndominiums, which are two-purpose buildings and serve as residences and business places at the same time.

Pros of Lofts and multi-purpose rooms.

Lofts and multi-purpose rooms are very handy in designing the floor plan of a barndominium, given the number of merits that each space allows. The spaces give homeowners a chance to maximize their square footage while keeping an open and airy feel. Lofts, on their part, add functionality to a home without the need for more land or building expenses. While multi-purpose rooms offer flexibility and adaptability to ensure the space can change workable according to the requirements of the person.

These design elements further make barndominiums a more sustainable and cost-effective option. You can reduce extra construction by making maximum use of vertical and open space, thus keeping the costs down and minimizing your ecological footprint. This flexibility in multi-use rooms lets you get more use out of each square foot, further enhancing the value of your home.

Outdoor Living Spaces: Wraparound Porches or Patios

From a Barndominium-thoughtfully fusing the rustic appeal of barns with the convenience and luxury of residential life-outdoor living space has been becoming more and more indispensable for modern home designs. Where utility and aesthetics are considered, few features can rival a wraparound porch or patio in enhancing the functionality and aesthetic beauty of a barndominium.

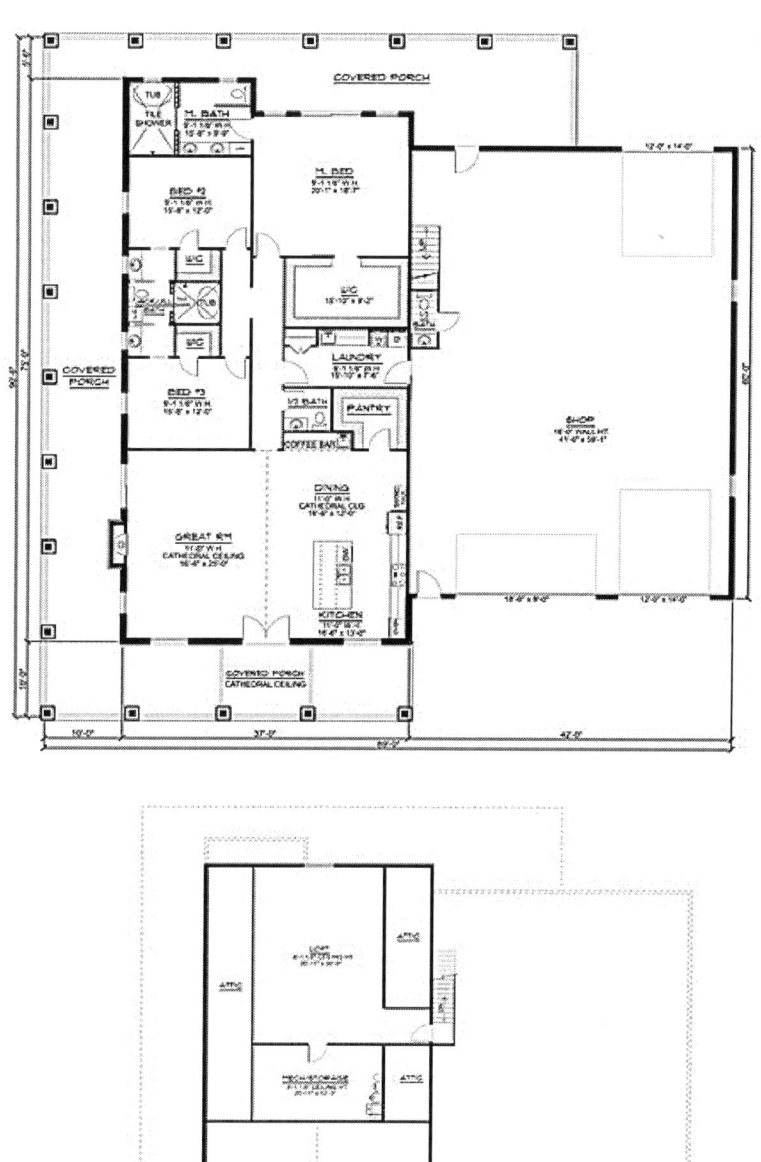

LOFT PLAN

It can also be one of the most attractive features. The success of either option will create a unified transition from inside out, or vice versa, adding to the overall cosmetic improvement of a home's aesthetic appeal. The section hereafter will touch on incorporating wraparound porches or patios into the barndominium floor designs, along with pros, design ideas, and concerns with doing such.

Attributes that Make Outdoor Living Spaces Attractive

The growing trend of homeowners to break out from the traditional boundaries that walls create has made outdoor living spaces hipper and more desirable. By using such space, one is exposed to the joys of nature, hosts guests, or sets up a mood for resting.

A well-designed outdoor living space becomes an integral part of the house, whether it's for sipping a cup of coffee in the morning, relaxing in the evening, or enjoying activities outside. In the case of a barndominium, the outdoor living spaces patios, or porches that wrap around the entire house complete the rustic and country appeal in these one-of-a-kind dwellings.

Why Consider a Wraparound Porch?

By the name itself, a wraparound porch is a porch that extends around several sides of the barndominium to encompass plenty of outdoor space with panoramic views all at once. Such a style of porch is highly favored among homeowners, especially in rural areas wherein the splendor of natural surroundings can be experienced. Installing a wraparound porch in a barndominium would be quite advantageous in several ways; these are some of them.

- **Maximized outdoor enjoyment:** With a wraparound porch, you will have outdoor living on all sides of the house. This, in return, allows you to take full advantage of the outdoors. If you want to sit and see the beautiful sunrise or see the sunsets, a wraparound porch gives you the perfect time to enjoy different views and sun exposure throughout the day.
- **Increased aesthetic appeal:** The wraparound porch adds aesthetic appeal to an entire barndominium. It makes the property feel more inviting and roomy, adding a touch of the South. Furthermore, the vastness of the covered outdoor space will be a perfect complement to the tall, open floor plans typical of barndominiums, thereby making the property feel even grander.
- **Extra comfort and shade:** Depending on the direction a house faces, a wraparound porch can add a little comfort and shade. The feature can offer protection from the elements, thus making outdoor spaces more usable if the weather is hot or rainy. Lower temperatures within the home can also be assisted by such a feature since shade is provided to the external walls. This means the home will require less cooling to maintain a comfortable temperature.
- **Seamless transition between indoor and outdoor spaces**: It may elicit a seamless transition between indoor and outdoor spaces with its occupants, with ease of flow from indoors to the outdoors. The openness of the house allows for airy sensations that are ideal for people who love staying indoors just as much as outdoors. There are many doors leading out onto the porch.
- **Value addition:** Apart from providing comfort and style, a wraparound porch increases the value of your barndominium. The innovation not only enhances the curb appeal of

your house but also the outside space, which becomes really utilitarian and thus quite attractive to any future potential buyer.

Wraparound Porches: Design Considerations

Design considerations that can be explored in integrating a wraparound porch into the barndominium floor plan include but are not limited to the following:

- **Structural integrity**: It's important to make sure the structural requirements of the barndominium are in concert with the porch design. The size of the porch must be in such a way that at no time will there be interference with the structure of the building in a manner that may obstruct natural light flow into the house.
- **Materials and design:** The type of materials used for the porch should blend well with the design of the whole barndominium. Barndominiums are known to possess a rustic character, which, of course, is reflected in these different materials that have been popular choices, including natural wood, metal, and stone. Let the materials be extremely durable and resistant to weather conditions outside.
- **Overall size**: The size of the wraparound porch should be in proportion to the overall size of the barndominium for which the structure has to be planned. In the case of a small-sized barndominium, the architecture might need to be more restrained to maintain balance, while in the case of a large barndominium, it can be more expansive.
- **Furniture and decor:** It is relevant to consider the furniture, plants, and other outdoor ornament arrangements that will be put onto the porch to create harmony in the living outdoors. To give a feeling of leisure, couches, swings, or hammocks that are comfortable to sit in can be provided. Additionally, outdoor dining areas provide practicality in terms of entertaining guests.

Why Choose a Patio Design?

With barndominiums, there is also an option for building patios, which are nice outside living areas. This might be a little comparative, but for the most part, patios are open-air areas that are always at ground level and can be constructed from things like stone, brick, or concrete. That is as opposed to wraparound porches, which are more intended to be raised and attached to the home. Here are only some of the many reasons why patios make a fantastic option as floor layouts for barndominiums:

- **Versatility:** Patios can be one of the most versatile spaces because they are quite accommodating in design and layout. Depending on how the barndominium was set up, they might be built in the backyard, a side of the house, or even a central courtyard. Because of this, patios are relatively easy to build and adapt to suit a wide range of tastes and needs for outdoor use.

- **Budget-friendly option:** Generally speaking, it is cheaper to install a patio than to construct a wraparound porch design. The patio requires fewer changes in structure to a house and is, therefore, cheaper for open-air living, especially for those homeowners who happen to be on a tight budget.
- **Entertaining in the great outdoors:** Patios lend themselves to outdoor entertaining and dining. Add an outdoor kitchen, a fire pit, or a barbecue area, and one has the perfect setting for such entertaining gatherings with extended family and friends.
- **Natural linkage to the landscape**: A patio creates a natural linkage to the landscape outside the walls, which is also known as natural integration. One can employ various landscaping elements such as plants, flowers, and water features to create a patio that perfectly merges with its surroundings. This will eventually produce a serene outdoor oasis.

Considerations to Make in Patio Designs

Some of the considerations when adding a patio to a barndominium floor plan include:

- **Size and placement:** First, consider the location and the space the patio is to occupy before making any decision on where to place it. It needs to face a location where outdoor vistas can be enjoyed while at the same time allowing space for seats and other elements like a garden or a fire pit.
- **Materials:** It must be focused on choosing materials that are complementary to the exterior of the barndominium and that consider the natural environment in which the construction will be located. Among the most adequate options that go by this type of rustic barndominium are the possibilities of stone or concrete due to resistance and little care needed.
- **Shade and comfort:** The inclusion of features such as pergolas, umbrellas, and even awnings can offer shade during hot days. As a result, the patio will be more functional under various types of weather conditions.

Garages and workshops: more room for extra vehicles and hobbies

In their most basic form, barndominiums intend to provide residents with open and roomy inside living space apart from the added advantages of sizeable garages or workshop rooms. Their aesthetic appeal is normally based on designs that are rustic and industrial; though internally, this can be tailored inside to be as luxurious or simplistic as one wishes. Such constructions are very often built with a metal frame, and they boast large, open spaces, which allows incorporating sizeable garages and workshops into the buildings' floor plans quite easily.

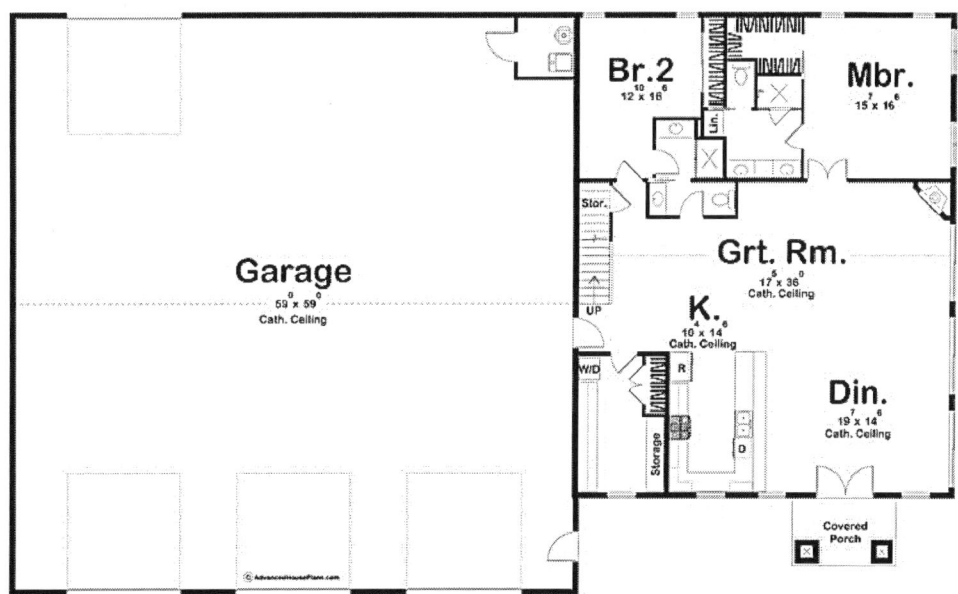

Barndominiums are attractive to many people due to the low maintenance involved, the overall affordability, and the wear and tear resistance of this type of building. Many barndominiums are constructed with steel framing and metal exteriors that greatly protect against the elements and vermin, hence reducing long-term maintenance costs. Barndominiums, other than that, have flexible floor plans to accommodate various requirements. These could include a large living area, extra bedrooms, or the main focus of this article: space for more vehicles and hobbies.

Why You May Need Extra Garage Space

Garage and workshop space is one of the most important planning aspects to consider when weighing the options to purchase a barndominium. This will be important for those homeowners who have large collections of vehicles like antique cars, motorcycles, boats, or RVs. It may also require a certain place where they can devote time to their hobby. With the ever-growing popularity of restoring automobiles, metalworking, and woodworking as hobbies, many a homeowner would surely love to have a huge workshop, stocked with all of the tools anyone could need or want.

If you have a barndominium, then the garage or workshop size is customizable according to your needs. Many floor designs incorporate huge garages, accommodating up to two, three, or more automobiles. Barndominiums, with their high ceilings, allow for the installation of vehicle lifts. Thus, they can be a perfect choice for automobile enthusiasts who may want to work on their vehicles in a professionally appropriate facility without leaving the convenience of their property.

Barndominiums allow personalized workshops to be designed that can be well-ventilated, well-lit and adapted to specific demands. This is great for the person whose hobbies call for a large workstation, such as in carpentry or craftwork. With due attention to design, such workshops can be designed with inbuilt storage facilities for tools and supplies along with workbenches. This would ensure that the area is not only vast but practical.

Changing the Appearance of Your Garage and Workshop

The biggest advantages of building a barndominium are the huge design freedoms. In contrast to traditional homes that may sometimes arrive with set floor designs, barndominiums can be constructed to the exact specifications that you may need. You will have the freedom to decide how much space you will use for your garage or workshop and how you want it designed.

Do you need space for five cars, a boat, and an RV, or are you looking for a simple garage that will fit two vehicles? Do you need a work area for woodwork with specialized tools, or do you want a place to store classic cars? The options are virtually endless. Many barndominium builders have lots of available floor plans that can be customized according to your preferences in terms of style and size, whether for a garage, a workshop, or any other utility such as heating and ventilation, down to plumbing.

For those needing even more tailored touches, the installation of garage doors, larger entryways, and even climate control can make all the difference. If you are in a cold region and must have some assurance that your working area can be heated, or if you have sensitive equipment or cars in storage, which need temperature regulation, the barndominiums can be adapted to such needs on your behalf.

Multi-Use is Not Constricted to Traditional Garage

Many barndominiums have garages and workshops, but most of the time, they are used for purposes other than automobile parking. They tend to be multi-purpose, so a person might use them as an office, studio, gym, or even entertainment area. They can easily accommodate the shifting requirements as time goes on due to how adaptable the floor layouts are in addition to how much space is provided.

For example, a car enthusiast may start by having a garage solely to store and take care of automobiles. They may choose, over time, to add workbenches, an office that will be used exclusively for the 'bookwork' associated with restoring automobiles, or even a reclining room in which to watch car shows after some time has passed. A Craftsman can have a woodworking studio in one part of the workshop, another piece kept for a station in metalworking, and one more section for storage and display of finished goods.

All you need is just a glimpse of your imagination to come to the view that these garages and workshops can also be used as family and friends' gathering places. For instance, you could grill outside and leave the garage door open so that it is easy to pass back and forth. If you are not currently using the workshop, such a garage may double as a home gym or yoga studio.

Popular Barndominium Floor Plans for Workshops and Garages

Barndominium floor plans are very popular to be used in garages and workshops. Barn floor plans are very popular, with quite a few specifically made for individuals needing a great deal of space for their garage or workshop. Some of the more common of these designs incorporate a great-sized garage that opens onto the living quarters and has separate entrances for the living quarters as well as to the garage itself. In many instances when these designs are brought into reality, that garage is large enough to fit a few automobiles in with further space for storage or workplace.

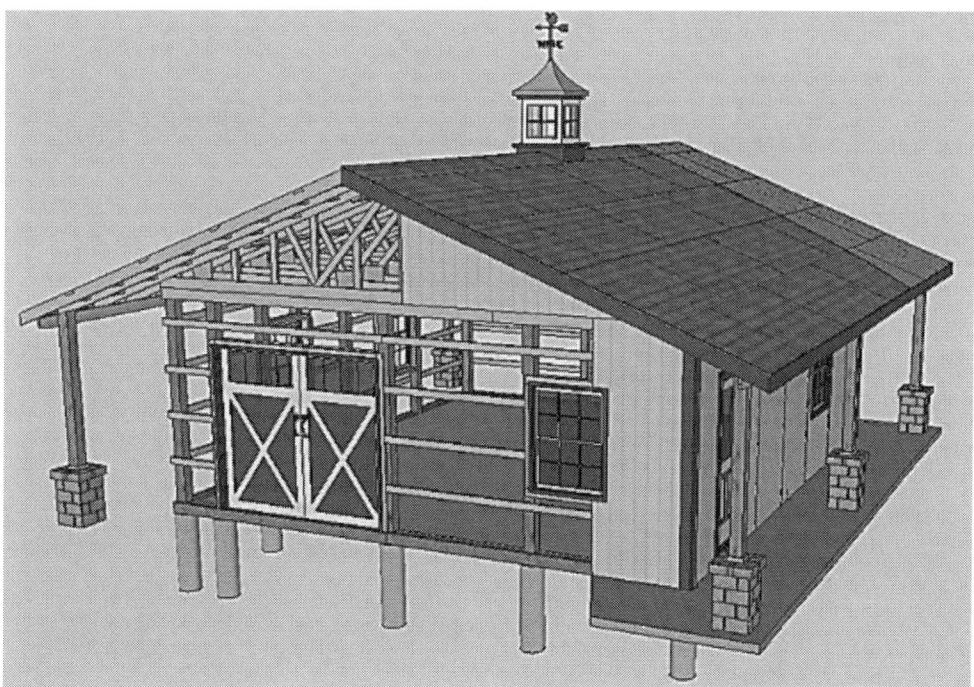

Another common setting is the "side-by-side" arrangement, consisting of having the garage or workshop situated right next to the living space, but separated by a breezeway or enclosed corridor. For those who want fast access from their house to their workshop but still desire some separation between the living and working rooms, this will be an excellent plan.

Of course, for those of you who need even more space, two-story barndominiums are another option. The living quarters are found upstairs, while the garage or workshop is downstairs on the lower floor. This would be perfect for those who need a large amount of garage space but want to maximize how much living space they have without sacrificing footprint size.

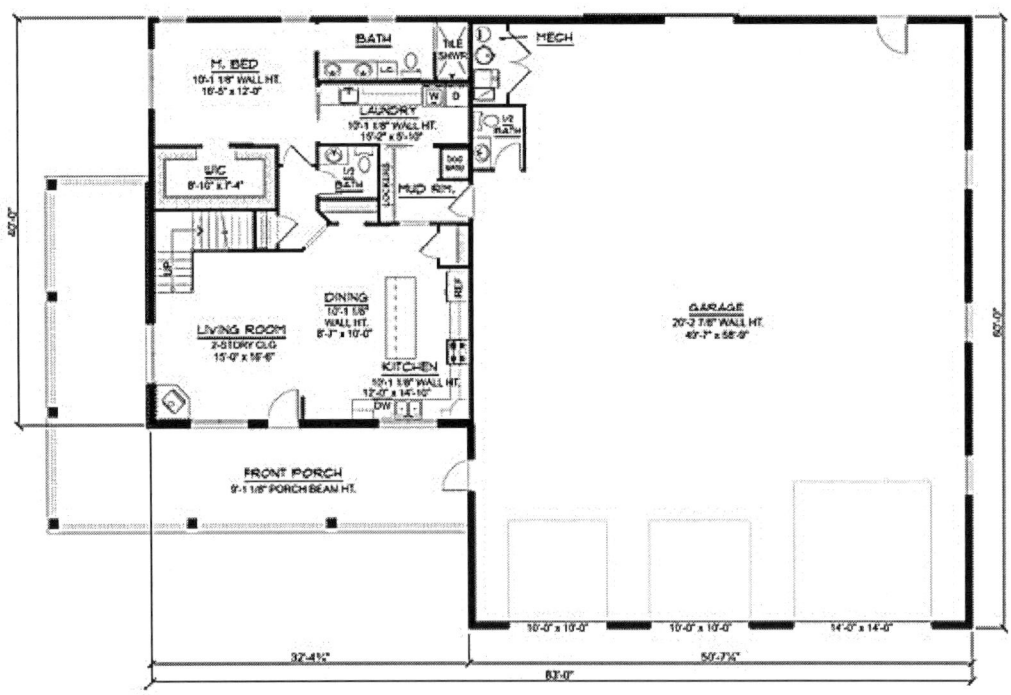

CHAPTER FOUR

Types of Barndominium Floor Plans

Small: 1,000-1,500 sq. ft; 1-2 bedrooms

A. 2 bedrooms, 1 bath barnominium (1,460 sq. feet)

Floor plan

- **Main floor**

About this Plan

This 2-bedroom, 1 bath barn style home marries rustic charm with modern living. Spacious enough with 1,460 square feet of heated living area, this home has been intelligently designed for comfort and practicality. For families preferring one-story living, the convenience of a single-story plan maximizes living without sacrificing the open invitation of living space.

Heated Space

- Total Heated Area: 1,460 sq. ft.
- First Floor: 1,460 sq. ft.

Unheated Space

- Garage Area: 940 sq. ft., accommodating 1 car

Floor Details

- Floors: 1
- Bedrooms: 2
- Bathrooms: 1
- Garages: 1 car
- Building Details
- Width: 40 ft.
- Depth: 60 ft.

Foundation Options

- Slab Foundation
- Crawlspace Foundation
- Basement Foundation
- Walkout Basement Foundation
- Main Roof Pitch: 6:12
- Exterior Framing: 2x6 Wood Framing

Ceiling Heights

- First Floor Ceiling Height: 14 feet

From open floor plans to many multi-foundation options, depending on the landscape, spaciousness, and practicality have been combined into one. The high ceilings create openness while the barn architectural style brings the timeless aesthetic appeal of the building. Whether one wants a serene country retreat or a more modern home infused with rustic elements, this plan certainly offers a wide base of flexibility for your needs.

B. Unique Accommodating Barn Home

This large 2-bed, 2-bath barn-style home melted together seamlessly the rustic charm of old into today's modern amenities. All on one single, accessible floor, the open and thoughtful layout offers 1,408 square feet of heated living space.

Livability Summary:

- Total Heated Area: 1,408 sq. ft., offering ample room for comfortable living
- First Floor: 1,408 sq. ft. of well-planned open-plan living.

Outdoor Living: Step outside and enjoy the generous porch and patio areas totaling a seemingly endless 1,920 square feet. Continue to provide endless opportunities for outdoor gatherings, relaxation, or simply to take in the beauty that surrounds you.

Structural Features:

- Floors: 1
- Bedrooms: 2-private and comfortable
- Bathrooms: 2-convenient and modern
- Building Width: 52 ft.
- Building Depth: 64 ft.
- Height: Soaring to 26 ft. high to create an airy, light-filled interior space.
- Foundation Options: A versatile design for a slab, crawlspace, basement, walkout basement-wherever the need or site conditions dictate.

Architectural Details:

- Main roof pitch: 8:12 for a classic barn-like look
- Exterior framing: Constructed with strong, resilient 2x6 to make sure it lasts a lifetime and is energy-efficient

Ceiling Heights:

- First Floor: Major living areas foster the illusion of size and openness with their dramatic 10-foot ceiling height.

With this barn house plan, you have achieved flexibility and style that will just make it perfect for those who are looking to meld indoor and outdoor living in timeless, rustic design.

Floor Plans

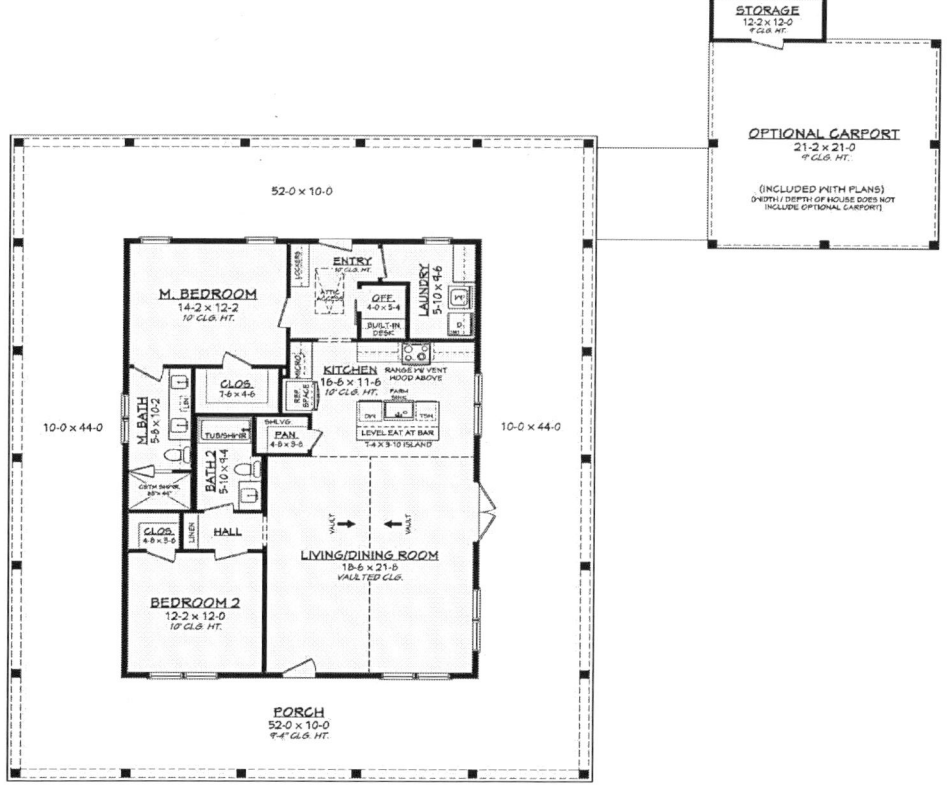

STORAGE
12-2 x 12-0
9' CLG. HT.

OPTIONAL CARPORT
21-2 x 21-0
9' CLG. HT.

(INCLUDED WITH PLANS)
(WIDTH / DEPTH OF HOUSE DOES NOT
INCLUDE OPTIONAL CARPORT)

52-0 x 10-0

ENTRY
10' CLG. HT.

M. BEDROOM
14-2 x 12-2
10' CLG. HT.

OFF.
4-0 x 5-4

LAUNDRY
5-10 x 4-6

BUILT-IN
DESK

CLOS.
7-6 x 4-6

KITCHEN
16-6 x 11-6
10' CLG. HT.

RANGE W/ VENT
HOOD ABOVE

M. BATH
5-8 x 10-2

10-0 x 44-0

TUB/SHWR

PAN.
4-6 x 3-6

LEVEL EAT AT BAR
7-4 x 3-10 ISLAND

10-0 x 44-0

BATH 2
5-10 x 4-4

CLOS.
4-6 x 3-6

HALL

LIVING/DINING ROOM
18-6 x 21-8
VAULTED CLG.

BEDROOM 2
12-2 x 12-0
10' CLG. HT.

PORCH
52-0 x 10-0
9'-4" CLG. HT.

Main Floor/Basement Stairs

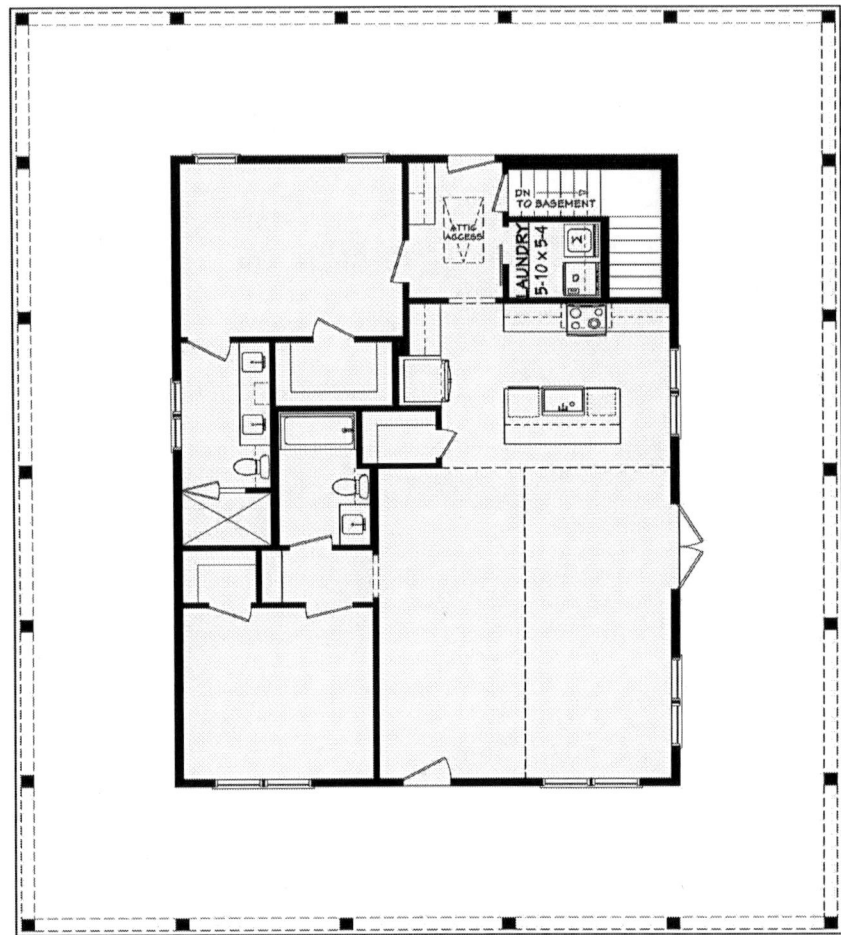

C. 2-bedroom, 2-bath barn

About this Plan

This 2-bedroom, 2-bath barn house contains all appeal with just the right amount of rustic charm and modern conveniences combined. This home seamlessly merges an uncompromisingly thoughtful 1,292-square-foot living space with a cozy, inviting atmosphere into a comfortable lifestyle it for a weekend retreat, a primary residence, or as a versatile secondary home.

Living Space and Layout

This is a single-story home that features 1,292 heated square feet on the main level. An open layout creates great flow and function throughout this home. With the number and size of the bedrooms and two full bathrooms, this design allows for ample personal space while the living areas remain central and connected.

Garage and Unheated Space

For the car enthusiast, or for one that would need additional storage space, the plan provides an additional 1,292 square feet for a 3-car garage. That is one extended garage space, always welcome, to be used at one's discretion for cars, a workshop, or even as a play area.

Building and Exterior Details

- Dimensions: Width-68 feet, depth-52 feet, and a total height of 25 feet are the magic figures that give this home an imposing yet inviting barn-like appearance.
- Roof: A roof with a 10:12 pitch gives style and practicality to the house, while 2x6 durable exterior framing lends it strength and insulation.
- Foundation Options: From flat lots to sloping sites this design is offered with several different foundation options. With a crawlspace, basement, monolithic slab, floating slab, and a walkout basement foundation, the options are vast, and you decide on what works for you or what your site calls for.

Interior Ceilings

Inside, the home is very spacious, with 10-foot ceilings on the first floor, which contribute to an open and airy design. The tall ceilings come together in a comfortable marriage with the barn-style architecture, feeling very grand but warm and intimate at the same time.

It was designed from functional layout to spacious garage to flexibility in foundation options for wide-range lifestyles in an open and inviting manner.

Main Floor

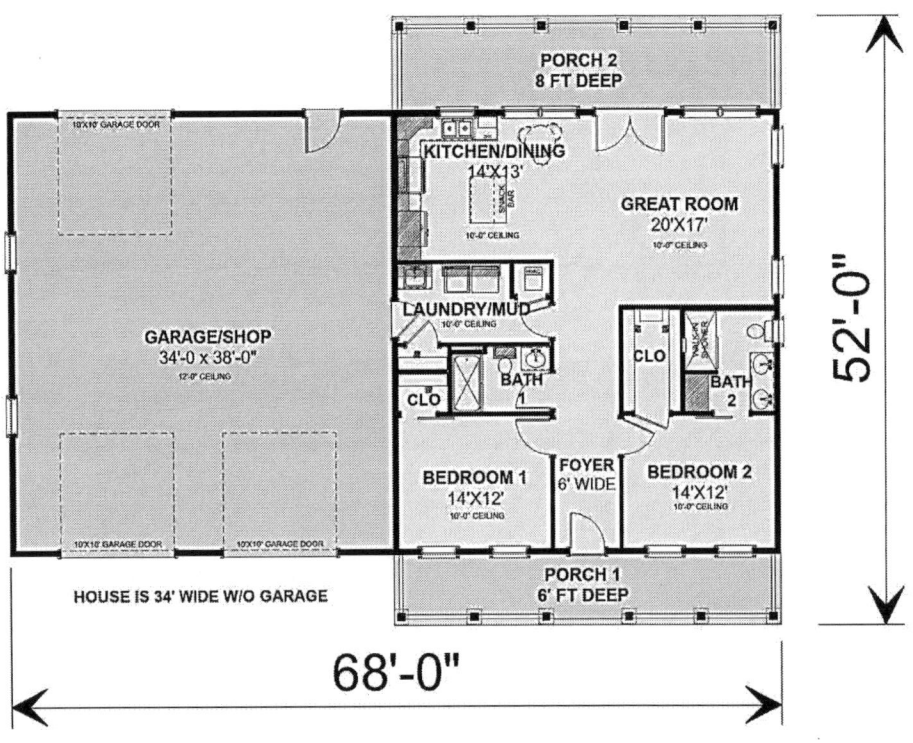

Overhead Floor Plan

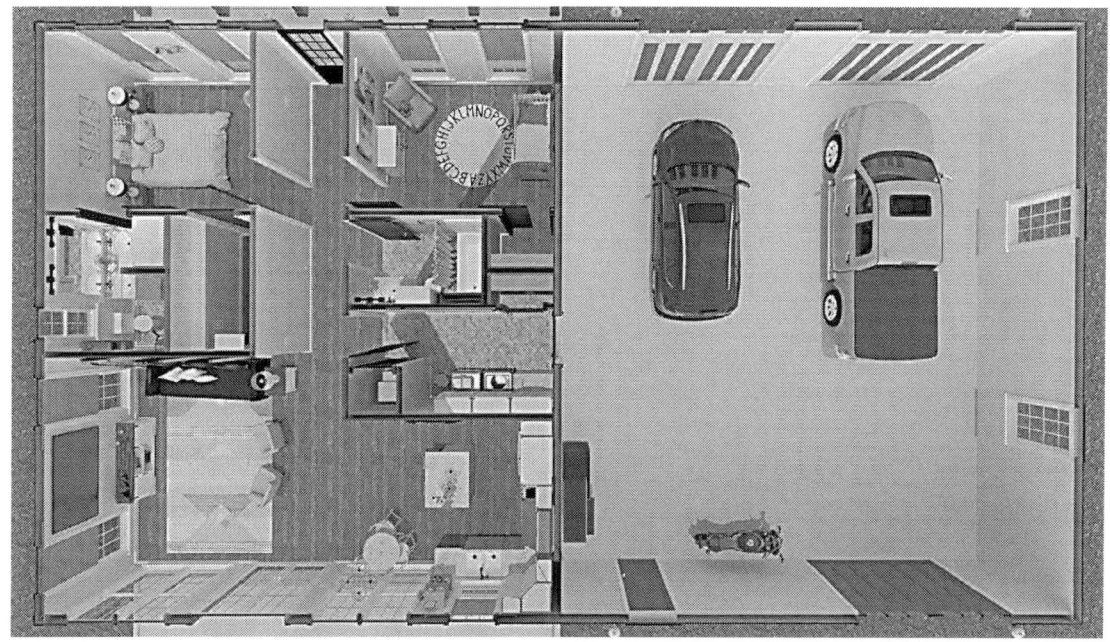

Mid-Size: 1,500-2,500 sq. ft;3-4 bedrooms

A. Huge 1800-square-foot barndominium with a 4-car, side-entry garage that skillfully balances modern living with functionality.

Welcome into this unique and increasingly popular barndominium lifestyle within this expansive 1,800-square-foot home. The floor plan is designed to strike a perfect balance between work, leisure, and practical living needs. With three large bedrooms and an open-concept living area, this is your sanctuary to find that perfect blend of relaxation and functionality while being spacious enough to live and work at home.

This property's crowning glory is the impressive 1,320-square-foot side-entry garage, which is large enough for four vehicles. The garage boasts four oversized 8' by 8' overhead doors, providing not only vehicle storage but also ample dedicated space for mechanicals, tools, and equipment for a variety of practical uses and recreation alike. Other thoughtful design touches include a door to a covered back porch, perfect for al fresco entertaining, and a secondary entrance leading directly to a well-appointed mudroom. A half-bath and built-in lockers make the mudroom a practical entry point for active families.

Enter through the front door into a warm living room with a cathedral ceiling opening and soaring above. A comfortable living area flows into an eat-in kitchen, perfect for everyday living or entertaining guests.

The master bedroom is privately situated on one side of the house and will serve as a tranquil retreat with lots of natural light. On the other side of the house, two other spacious bedrooms will offer comfortable accommodation for family members or guests, offering privacy and space for all.

Designed for aesthetics and durability, the exterior is cladded with corrugated metal siding, fully capturing the rugged, industrial charm of a barndominium. This gives a sleek and durable finish, including longevity with low maintenance, and gives a house a characteristic modern appeal that perfectly fits into both rural and suburban settings.

Floor plan

- **Main floor**

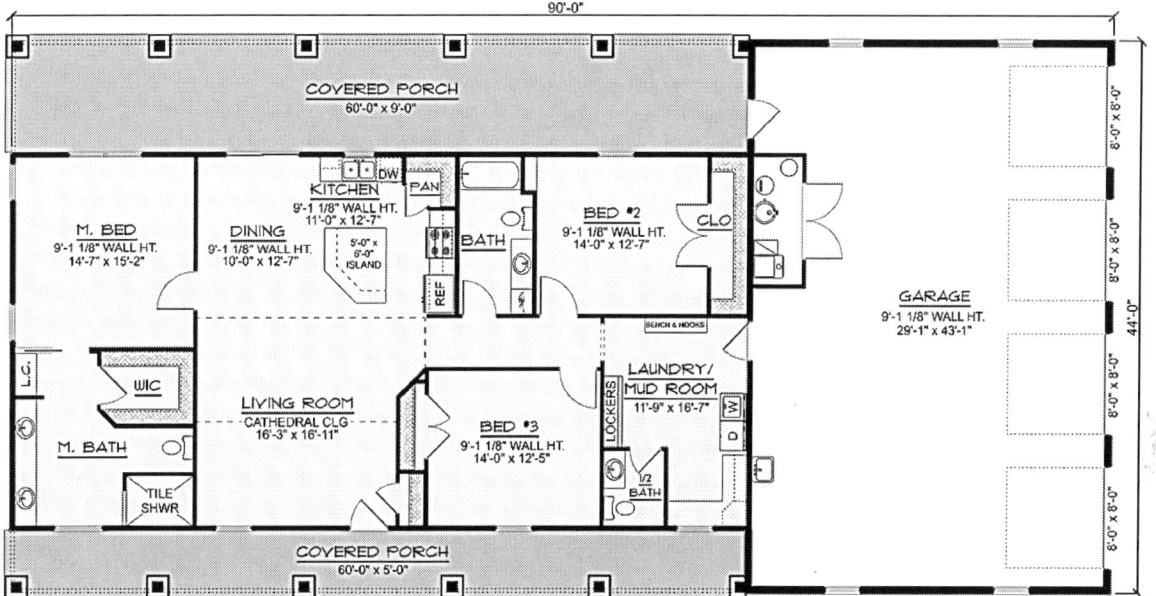

Main Level Basement Option

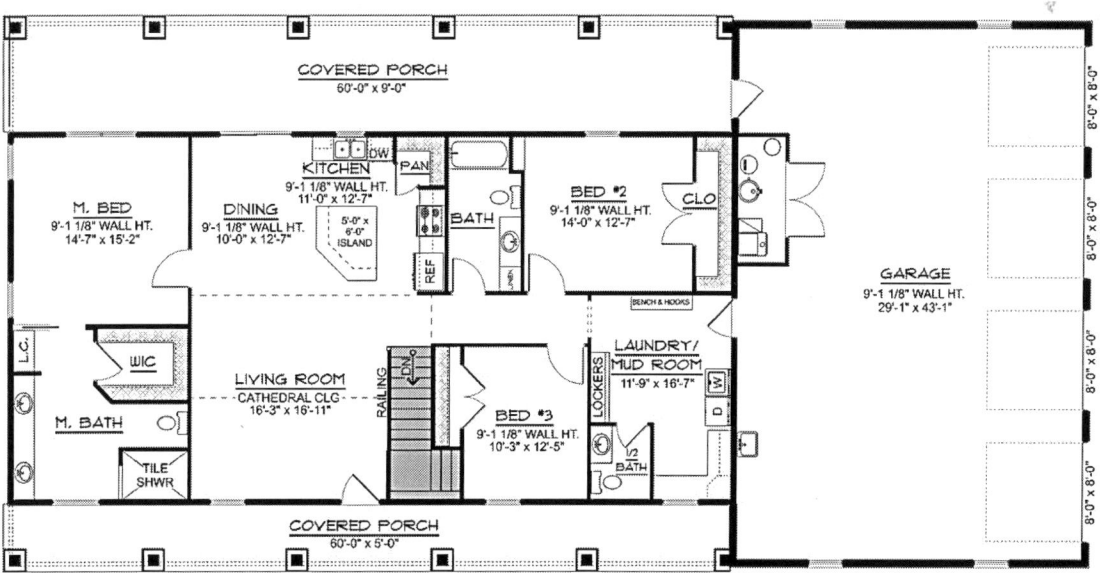

B. Open 2-bedroom barndominium style house plan with flexibility in living areas, and a drive-through garage.

This thoughtfully designed barndominium home combines functionality and style in just the right way to wrap up 1,578 heated square feet. The lower level includes 219 sq. ft., which is a versatile storage room or office along with a full bath. Upstairs living gives way to 1,369 sq. ft. of open-concept living, including two cozy bedrooms, a vaulted living room, and a well-appointed kitchen that enhances the expansive feeling of the home.

This would give any car enthusiast or hobbyist 1,499 square feet of parking space on the ground level. The garage includes two overhead doors at the front, each 10' by 8' to allow easy access into an open area without obstruction. The left bay features a third overhead door at the back, also measuring 10' by 8', to allow this very rare and practical drive-through feature for added convenience. Whether projects in the garage or outdoor playtime, a full first-floor bathroom is conveniently accessed from outside to clean up before heading upstairs.

But this home also offers 380 square feet of covered patio space in addition to its functional interior spaces, provided by overhanging two-car bays protecting this outdoor area from the elements. The upper level extends outdoor living with its 380 square foot deck accessible from the kitchen and perfect for entertaining or enjoying the view. Adding further to the charm and utility of the house's outdoors is a covered porch of some 240 square feet, accessed from the living room via some elegant French doors.

This will work nicely as a guest house or an apartment on a larger property, or it's comfortable and practical enough to serve as a place to live while building your primary residence. From rustic charm to modern convenience, versatility in space can be tailored to fit a wide array of uses and lifestyles.

Floor plan

- **Main level**

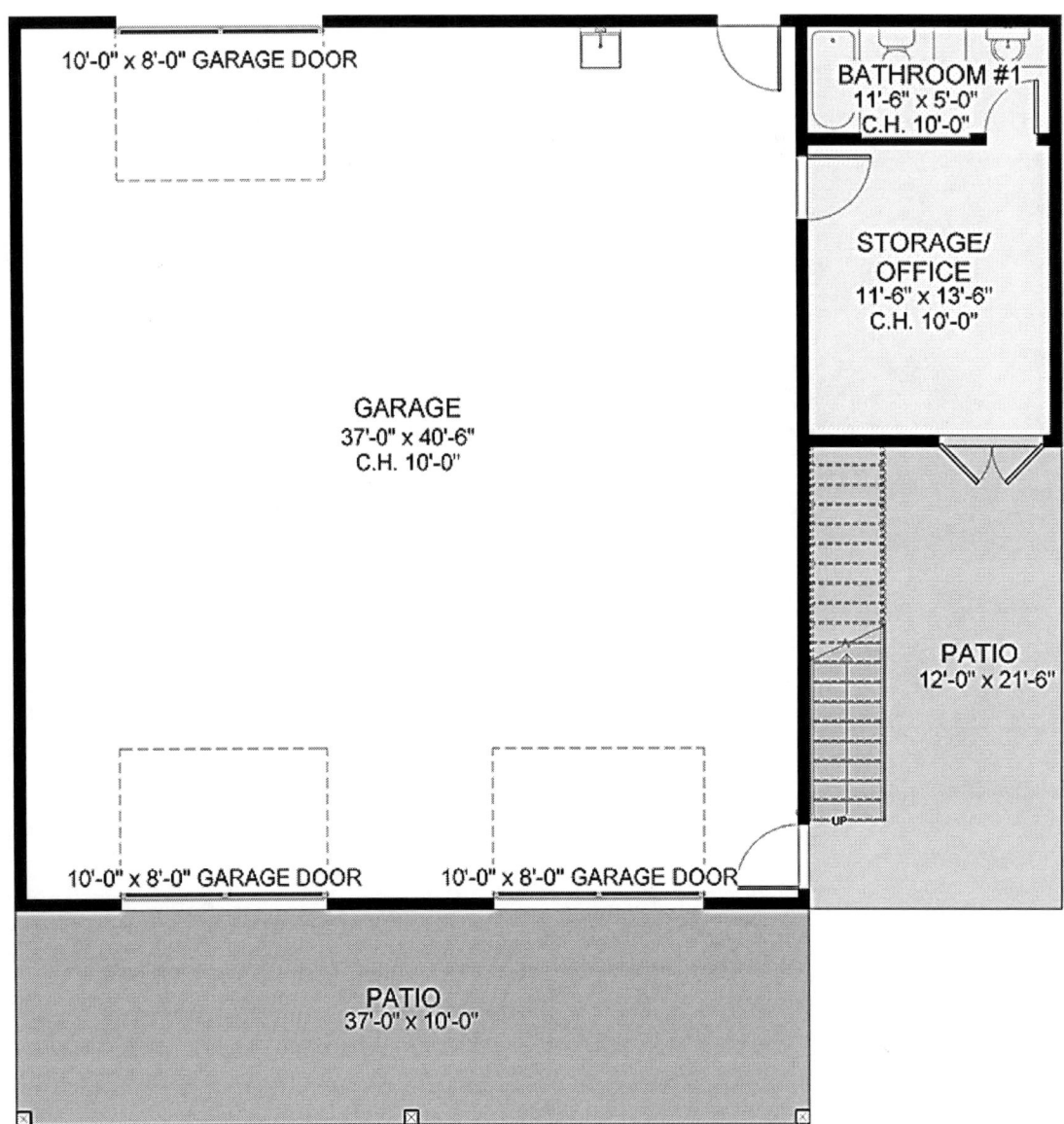

Second Floor

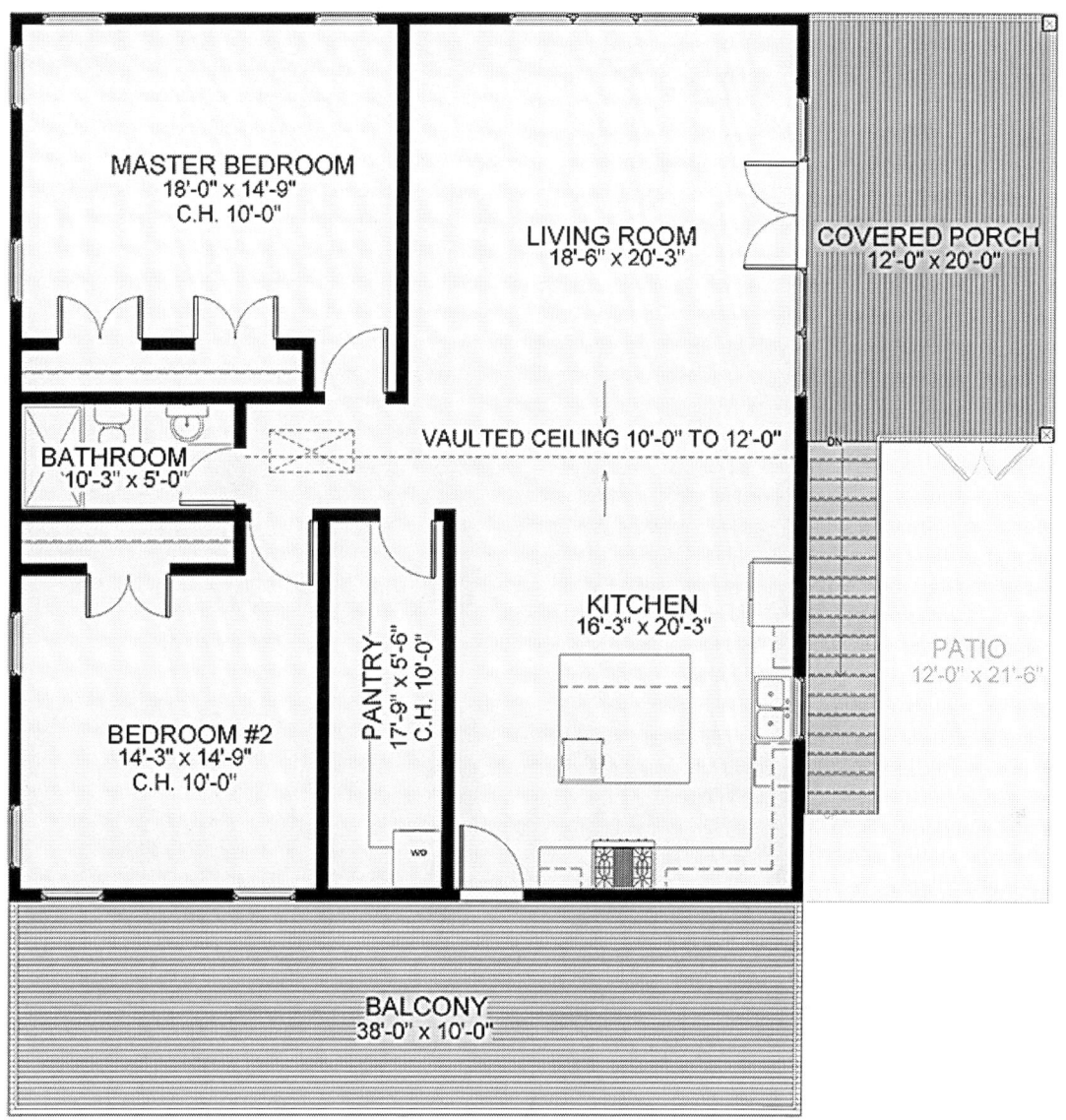

MASTER BEDROOM
18'-0" x 14'-9"
C.H. 10'-0"

LIVING ROOM
18'-6" x 20'-3"

COVERED PORCH
12'-0" x 20'-0"

BATHROOM
10'-3" x 5'-0"

VAULTED CEILING 10'-0" TO 12'-0"

PANTRY
17'-9" x 5'-6"
C.H. 10'-0"

KITCHEN
16'-3" x 20'-3"

PATIO
12'-0" x 21'-6"

BEDROOM #2
14'-3" x 14'-9"
C.H. 10'-0"

WD

BALCONY
38'-0" x 10'-0"

C. Stunning barndominium with large garage and inviting living spaces

If one seeks a house that easily and comfortably marries spacious living with exceptional storage options, this sprawling 2,016 sq. ft. barndominium could be just the right match. Boasting an impressive five-car garage, this home will satisfy any need for vehicle space, hobbies, or additional storage, opening directly onto a covered rear porch for ease of outdoor entertaining.

Inside, it opens into a spacious open-concept living area ideal for entertainment and socialization, or quiet hours together as a family. The L-shaped kitchen is the delight of any chef, with a walk-in pantry for ample storage and a center preparation island where meals are prepared or casually dined on. The dining area sits adjacent to the rear porch via sleek sliding glass doors, which serve to flow the indoors out and are perfect for morning tranquility or evening gatherings.

This thoughtful design continues down the hall to two well-appointed guest bedrooms with comfort and privacy, plus a spacious primary suite designed to be a personal retreat. The primary bedroom boasts luxurious amenities and ample space, making this home's function and style come full circle in perfect harmony.

This is less a house than a thoughtfully imagined refuge marrying modern convenience, ample storage, and welcoming ambiance in equal parts for daily living and entertainment.

Floor plan

Main level

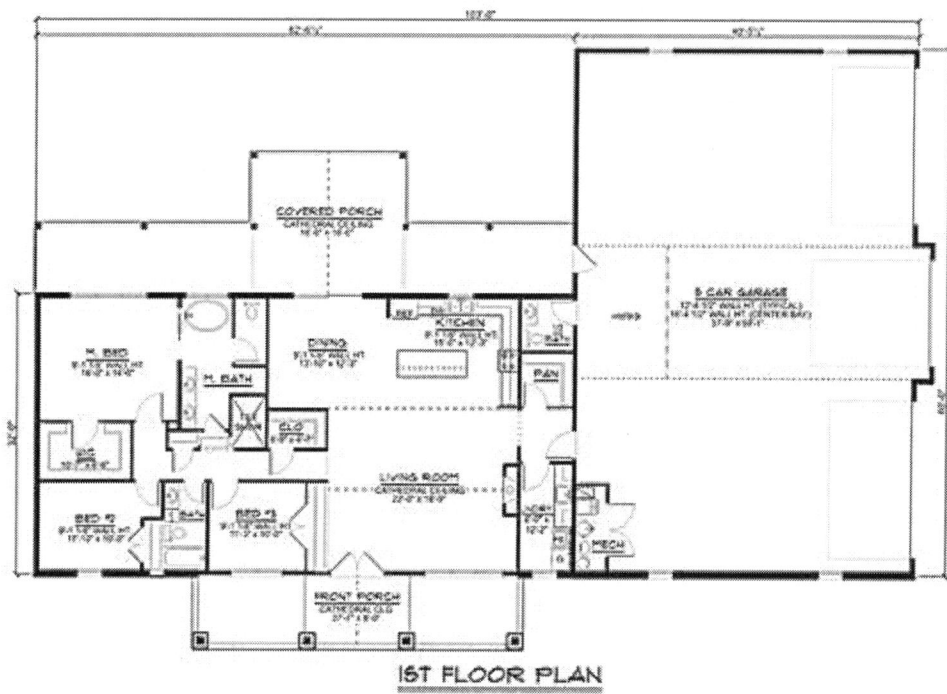

IST FLOOR PLAN

Large: 2,500+ sq. ft; 4+ bedrooms

A. Big garage plus wraparound

How is this perfect? Introducing the perfect 2,500-square-foot barndominium that seamlessly melds rustic charm with modern farmhouse style. The beautiful, wood-framed structure is accented by a very spacious wraparound porch perfect for outdoor living and entertainment. Perhaps one of the most striking features of this property would have to be the enormous three-car garage, which will afford plenty of room for vehicles, storage, or even a workshop.

The open-concept interior is bright and airy, enhanced by the high, vaulted ceiling and the stylish kitchen, which features a great big island for casual dining or food preparation. The walk-in pantry brings convenience along with storage, keeping all you need within easy reach.

A primary suite on the main level-most luxurious-finds its core in the home, revealing the very essence of a peaceful retreat with its spa-inspired bathroom to please even the ultimate in relaxation. Rustic elegance beams within the ceiling of the bedroom, making it cozy yet refined.

Combining the best of comfort, style, and functionality, this home design is a perfect place to call home.

Main floor

B. 4 bedrooms, 3-bathroom barndominium plan

Main floor

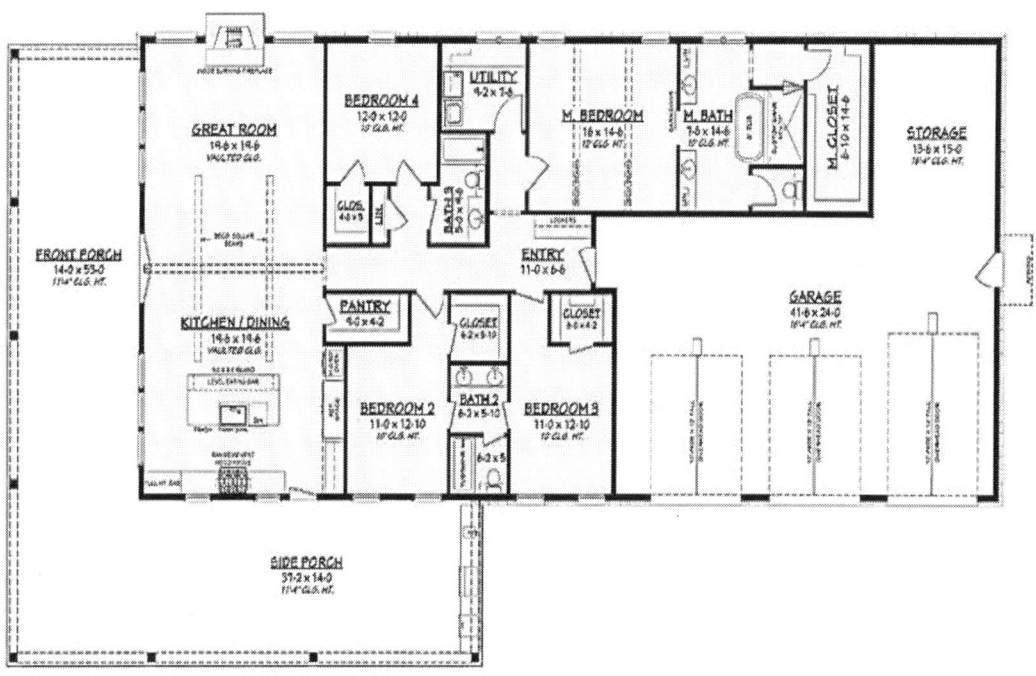

With a specific design to meld rustic charm with the luxury of modern space and entertainment, this grand barndominium surely does set the standard. The house has four bedrooms of generous size, three well-appointed bathrooms, and spacious walk-in closets throughout. The open-concept great room, dining area, and chef's kitchen make up the heart of the house, with their soaring vaulted ceilings creating a light, airy atmosphere that invites revelry.

Giant windows allow natural light into the living spaces, while the wraparound porch creates fluid indoor-outdoor living. The kitchen features an eat-at island, ample counter space, and a huge walk-in pantry to satisfy culinary needs. The oversized garage provides additional space for vehicles or projects, with the added convenience of additional storage above the main living areas-accessory storage perfect for seasonal items or hobbies.

CHAPTER FIVE

Making Your Barndominium Floor Plan Functional

Designing a barndominium should be functional. The elaborate floor plan ensures that each room in the house is beautiful and useful during daily living. First, there is a need to consider the flow of high-traffic areas: the kitchen, the living room, and the dining space. Having the openness of the concept design will enable the space to have an aspect of openness and navigate through elements with ease. Consider ways that best use natural light, adding to reducing reliance on artificial lighting during the day. This makes the house more luminous and inviting visually by the strategic location of the house's doors, also by installing large windows.

Other important features, regarding functionality, involve storage: take the clutter from your living spaces and have them look neat by availing yourself of built-in shelving, closets, and other types of storage solutions. With much larger barndominiums, it would be quite beneficial to the user that they should have areas of work, playing, and rest distinctly outlined for better equilibrium and comfort.

Consider future needs-particularly those people who require additional bedrooms or multi-purpose rooms that can be transformed as the years go by. You can make sure, by design, that your barndominium will be functional and able to adapt to whatever changed lifestyle might happen in the future by designing for flexibility every time a decision is made.

How to make your space functional: the number and placement of bedrooms, bathrooms, and office space.

Especially in the setting of a barndominium, designing a functional living space does require great care in emphasis on the number of bedrooms, bathrooms, and office space, along with the arrangement of all these.

With open floor plans, reasonable prices, and the ability to be customized, barndominiums are becoming very popular. While the outside of these structures is built so that it resembles a barn, the inside can be decorated in an extremely opulent manner or as simple as the owner would like.

A balance must be struck in optimizing the function and the layout of each room to make the space comfortable, functional, and appealing. To help make your barndominium a useful and pleasurable place, here is how you may plan the essential spaces.

A. The number of bedrooms and their locations

Among the key elements involved in developing a floor plan are the bedrooms, their number determined by the household size, frequency of guests sleeping over, and your lifestyle.

Bedroom Sizes

You need to know who will be occupying your barndominium. For many homes, each person in the household may want or need a bedroom to call their own. While a couple with children may require three to four bedrooms, an older retired couple may be satisfied with only one or two bedrooms. Of course, if you still foresee friends coming over regularly, then at least one guest bedroom or two will be nice.

Also, when one is considering the number of bedrooms, resale value in the future might be a prudent thing to consider. Generally, a house with more bedrooms enjoys greater market appeal, and therefore, selling it in the future will not be much of a problem. Because of this, a three-bedroom layout is ideal for the most part, since it is flexible without relying too heavily on space or finances.

Bedroom Placement

Even more crucial than the number of bedrooms is the orientation of the residence. The bedrooms must be placed in a fashion to provide a balance between accessibility and seclusion.

- **Primary Bedrooms**: The orientation of the master bedroom, or primary bedroom, is best if it is removed from high-traffic areas such as the living room or the kitchen to provide optimum levels of privacy and quietness. If the owners are older, or for convenience's sake, if they plan on staying in the house for an extended time, it is usually placed on the main level.

- **Secondary Bedrooms:** In homes with children or guests, the location of secondary bedrooms should reflect a balance between proximity and solitude. Wherever possible, locate them close to each other, but especially for small children who need to feel their parents nearby. In addition, the guest rooms should be constructed farther from the main bedroom to provide them and owners with more privacy.
- **Home Office Conversion**: If you think that the number of family members living in your barndominium will fluctuate over time, then you could already consider how you'll convert one or more bedrooms to a home office or hobby room later on. That way, you can sleep at night knowing it's going to be workable to adapt to the situation.

B. How many bathrooms should there be, and where should they be located?

Bathrooms are another functional necessity. The number of occupants the space will have, and how frequently the occupants will be there will determine the number of bathrooms and where they will be located.

Number of Bathrooms

Many experts suggest that a home of three or more persons deserves at least two bathrooms. This is to avoid the inconveniences caused and also the bottlenecks during the morning rush. In general,

- One bathroom would suffice if the members in the household were a few or just couples.
- The house, with two bathrooms, would be ideal for a family with two or three children.
- For larger households or those that entertain regularly, at least three bathrooms become a requirement.

The most common setting is one full bathroom being en-suite to the master bedroom, with another bathroom shared between the secondary bedrooms, and a half bathroom or powder room for visitors proximate to the primary living area.

Location of the Bathroom

The location of the restrooms in your barndominium is a big factor that determines the level of comfort and flow of everyday life. Among many, here is some practical guidance:

- **Master Bathrooms**: The master bathroom is to be installed en suite, directly attached to the principal bedroom. Double vanities, a soaking tub, and a walk-in shower are just several common elements included when designing rooms intended for luxury.
- **Secondary Bathrooms:** Shared bathrooms in secondary bedrooms may either be located between sleeping rooms that share a wall-the Jack-and-Jill arrangement or off a hall. This is less costly because the shared wall can save plumbing costs.
- **Guest Bathrooms:** Guest convenience dictates that a half-bathroom be located near the living spaces. It is an awkward feature for guests to have to enter private bedrooms to gain access to a restroom. A powder room placed near the kitchen or the living area is an attractive alternative to this problem.
- **Accessibility**: If your barndominium is designed with multiple stories, having at least one bathroom on each floor can be a huge convenience because it lessens how many times you have to go up and down the steps.

C. Office Space: A Must for the Modern Home

With working from home becoming so prevalent, it's becoming more of a priority to dedicate space in your house to a home office. You can add a layer of functionality to your barndominium with an office addition. This is whether you work from home full-time or just need a quiet place to handle your projects and bills.

Office Dimensions

You will have to size your workplace according to the hours that you will put in there, plus the equipment you are going to require. Perhaps you may just need a desk and chair for occasional usage, but if you work out of your home and this is your work headquarters, you are going to need a larger place with built-in storage, a big desk, and seating for customers or colleagues.

If you require a bigger workspace yet have only a little square footage, you can make it multi-purpose by combining it with a guest room or a hobby space.

Staffing of the Office

Productivity and privacy are directly proportional to the site where you house your work area. A few suggestions are:

- **Near the Master Bedroom:** You can place the office beside the master bedroom if you work late into the night or during odd hours. This allows you to get into the office easily while providing you with privacy. Ensure soundproofing is considered so other people are not disturbed.
- **Separated from Living Spaces:** It is located much farther away from the areas with main living. An office that is situated a considerable distance from where most of the main living occurs is ideal for people who need a quiet and focused environment in which to carry out their work. An office found in a converted bedroom or some obscure part of the house can offer the peace that is mainly needed in concentrating on one's job.

- **Natural Light:** This has been proven to elevate one's mood and raise one's level of productivity. When possible, place your workspace near the windows so you can enjoy the outside scenery while working.

D. **Flexibility and possibility for customization in the design of barndominiums**

Among many reasons, barndominiums are extremely popular because of how flexible they are. The open floor plans allow you to adjust the number of bedrooms, bathrooms, and offices with their locations to your specific needs. You'll have no problem making it serve a range of purposes for your growing and changing family. It could well be the case, for instance, that a home office, at some point in time, would perhaps be converted into a bedroom or that the guest room would then serve as an entertainment area.

With great contemplation on how many have and must-haves, where they go, and how much flexibility they allow, it will truly enable you to make your barnominium not only functional but comfortable and serve your needs for years to come.

Kitchen and living design: open versus traditional, islands, and storage.

Besides the fact that the modern home has become a utilitarian environment, it also reflects the personal style and tastes of the person and his trend-changing role in the world. Therefore, the question of how to choose between open and traditional floor plans can be found among the most contentious issues in any given design of living areas and kitchens.

The use of islands and innovative storage may also significantly affect the beauty and functionality of the space. Against the background of barndominium floor design and modern living, let's look into these from every perspective.

The most important decisions a homeowner will have to make in designing (an open layout versus a traditional layout.)

Perhaps one of the key decisions to be made when building both the kitchen and living areas in any home, especially a barndominium, is whether to go with an open floor plan versus a traditional floor plan. Barndominiums lend themselves to open floor plans as they usually have large interiors with high ceilings and sometimes a combination of rustic and modern design elements. However, there are some very real reasons to examine both options below.

- **Open Floor Plans**

In the past couple of decades, open floor plans have become increasingly popular among those who like the airiness and smooth flow from space to space. In an open floor plan, the kitchen, dining room, and living room often flow into one huge space with hardly any partitions separating it. Such a layout allows unobstructed interaction and visibility between areas of the house. This type of design is especially appealing to those who like to throw parties and other congregations among friends and family, as it builds a sense of socializing in the same space by cooking, eating, and relaxing.

This open layout, common in a barndominium that usually focuses on the feeling of space and volume, will maximize the architectural features of such a home, such as exposed beams, vaulted ceilings, and large windows. The openness adds to the feeling that the structure is open and makes even the smallest barndominiums appear larger and brighter than they are. It is also very appropriate for those who enjoy the minimalistic look of modern designs, which come with little clutter, allowing the light to flow.

Open floor plans are not without their share of problems. One of the major problems is that there are no well-defined areas. This can lend to an aspect of chaos should the design not be well thought out. This may not be a very suitable option for families that have children or for people who place a premium on quiet and separate spaces because noise and smells from the kitchen can easily spread out onto the living area.

- **Traditional Layouts**

By contrast, traditional floor designs are those that offer more separated floor spaces. The kitchen would usually be in a different area than the dining room and the living room in a traditional floor design. Each room would perform a specific purpose, which may have the effect of creating that personal feeling of warmth and closeness that some homeowners find very attractive.

Being an abode caused by a traditional barndominium plan, many zones could be housed in the house. Perhaps these could be one of the different advantages it has to offer, especially for those people seeking quiet places to retire away from maybe the hustle and bustle happening in the living room. It can also give one more control over the sound and odors emanating from the kitchen are not as apt to permeate the entire house. Those familiar with the inside of a barndominium know they are touted for their spaciousness and rustic appeal; using the right design, a classic plan has the potential to retain both attributes.

As such, traditional floor plans can seem to be much more closed-off and not utilize the open and spacious areas that a barndominium building commonly offers. When a family wants to have a point of congregation in their own home, the compartmentalizing of space can be considered a drawback.

How Kitchen Islands Factor in with the Design Process

The kitchen island has become standard in modern design, regardless of whether the homeowner decides on an open or traditional plan. The islands work well for barndominium homes, like the one depicted above, as they can add function, storage, and a focal point to the space.

- **Kitchen Islands for Open-Floor-Plan**

Under an open floor plan, sometimes a kitchen island can serve as a dividing line between the kitchen and the living or dining part of the house. This also aids in maintaining some semblance of separation without using walls, thus preserving the open flow but still offering a workstation that is useful nonetheless. In addition, islands provide extra counter space that simplifies meal preparation processes, especially when the household contains a large number or when guests are being entertained.

Besides, islands can be made just like barn-dominiums: a mix of modern and rustic designs. Adding a wide wooden island with metal accents can bring warmth and texture to the room, all the while it seamlessly blends into a combination of classic barn features along with contemporary finishes, such as stainless-steel equipment.

- **Kitchen Island for Traditional Layouts**

Even if the layout is much more segmented and traditional, the kitchen island is still going to be a profoundly practical part of it. It is, in particular, useful to kitchens that have less space due to additional counter and storage space. The island may be used in the traditional kitchen as an auxiliary breakfast area that is casual and used for breakfast or other quick meals without having to set up the dining room.

It is also possible for the island design in a traditional area to lean more toward the classic style. This could be achieved with ornate cabinetry, natural stone counters, and perhaps a more ornate finish that suits the enclosed aspect of the kitchen.

Barndominiums with Storage Solutions: Making the Most of Available Space

Storage is one problem known to be one of the most cumbersome design issues any home has to face. This again would be very understandable, in particular in a barndominium where the open floor plan can restrain as many built-in storage alternatives, especially in the kitchen and living rooms. This, again, is very much true in the case of barndominiums.

- **Open-Bedroom Layout Storage**

Storage solutions are very important in an open plan since they keep clutter at bay and maintain the feel of the space clean and open. Most people prefer installed cabinetry and drawers on their kitchen islands since they can provide ample facilities to store cookware, utensils, and small appliances. Open shelves can also be used to showcase exquisite dishware, pots, and pans that will add beauty to the kitchen with its open shelving.

Shelves and storage units can also be mounted on the walls so they meld in seamlessly into the living room. Books, electrical equipment, and knick-knacks can be stored out of the way in these without sacrificing any precious floor space.

- **Traditionally Designed Layout Storage**

Due to the private rooms typical of traditional layouts, there are typically more spaces for dedicated storage. You can optimize the pantry space available within the kitchen by adding in-built cupboards and shelving. Furniture with inbuilt storage space can also be availed for the living room. For example, a coffee table with some hidden compartments or a television unit with ample shelving can be a fine example of furniture doing this task.

Master Suite Design: En-suite Bathroom, Walk-in Closet, Private Outdoor Space

Designing the master suite is all about creating a beautiful, private room in your house personal place where one can retreat into their private sanctuary. People working with the barndominium floor plan have even wider options for building a vast and peculiar master suite. That is because smart homes boast open and flexible layouts, hence offering more scope for customizing.

An en-suite bathroom, a walk-in closet, and a private outdoor space are typically the three must-haves that feature in any well-planned master suite. Each of these features can be personalized to reflect the individual's preferences and the requirements of their lifestyle, as well as the style of the barndominium overall. In this section, we'll outline three such features, focusing on how they contribute to both the aesthetic and functional appeal of a barndominium master suite.

The en-suite bathroom: A perfect blend of luxuriousness and practicality

By definition, a modern or contempoarary master suite will include an attached bathroom. This provides seclusion but also extends the sensations of being in the bedroom and embraces an ambiance similar to that of a spa. In the context of the barndominium, which often features open spaces and rustic charm, the en suite can share this style or compete with it through more modern and sleek surfaces.

Key points in the En-Suite Bathroom

- **Freestanding Tub**: Amongst the en-suite bathroom features that make a house almost irresistible to homebuyers is an en-suite bathroom with a freestanding tub. Whether you're going ultra-modern or rustic farmhouse, a beautifully carved tub can be a showstopper in any bathroom. Putting a bathtub beside a picture window in a barndominium-a space defined by open, expansive floor plans- allows any occupants to enjoy the picturesque view of the countryside outside while indulging in relaxation.
- **Walk-in Shower:** For the person building the barndominium, it can be as big as possible and needs to have rainfall showerheads, benches, and even a few water outlets for a more luxurious experience. Large-format tile, natural stone, and even reclaimed wood walls can be used to give this area in the barndominium an air of rustic charm while keeping the space contemporary.
- **Double Vanity:** A double vanity allows for two people to have access to the bathroom at the same time without feeling congested, hence allowing mornings to run more smoothly and quickly. You can select from wood vanities, custom-made to extend the rustic look into the barndominium, or marble worktops and floating sinks to provide a modern contrast.
- **Natural Light**: A large-sized window or skylight can allow ample natural light. Generally speaking, in an en-suite bathroom, natural light is fairly important. Most barndominiums come with high ceilings, which can accommodate tall windows and let a lot of light into the bathroom. In this case, frosted glass or a carefully positioned window will be the best option for securing privacy in that space.
- **Heated Flooring**: Most barndominiums are constructed in rural or temperate areas; therefore, heated flooring adds luxury, particularly during the colder periods. Natural materials, such as stone or tile, go great with the rustic elements of the home, providing comfort.

The Walk-in Closet: An Elegantly Organized Space

A walk-in closet is not for storage alone; it could also be a part of the master suite, serving functionality or aesthetics, whichever one prefers. The well-planned walk-in closet makes the daily routine so much easier and adds luxury to your master suite. The wide-open layouts found in a barndominium create an ideal opportunity to create a truly customized walk-in closet that would suit specific storage needs and reflect the overall aesthetic of the home.

Design Elements of a Walk-in Closet

- **Custom Cabinetry**: Adding in custom cabinetry instantly raises the bar on walk-in closets. Barndominium homes, which frequently incorporate natural wood and rustic finishes, are the perfect canvas to add these elements in the form of wooden shelves, drawers, and cabinetry. These custom pieces can be designed from simple storage for shoes and handbags down to a place for everything.
- **Island or Seating Area**: Where space allows, an island in the center of the closet affords added storage for accessories, jewelry, and seasonal items. It also provides a surface on which to fold clothes or lay out a particular outfit. Some include a bench or a small seating area to wear shoes or enjoy the luxury.
- **Lighting**: A walk-in closet is not complete without good lighting. For its more rustic aesthetic, barndominiums may call for softer lighting. Place recessed and decorative lights like chandeliers, but also pendant lights, in your space. Placing LED light strips inside the cabinets can light up both the shelves and hanging parts so that all of the clothing and accessories are visible.
- **Full-Length Mirror**: No walk-in closet is complete without one. Position the mirror to reflect natural light streaming in from nearby windows or use it to make the space feel even larger. A wood-framed mirror in a complementary style to the rustic nature of the barndominium helps the closet fit into the overall design of the home.

Private Outdoor Space: Bringing the Outside In

One of the unique advantages of barndominium living is in the unified flow of indoor/outdoor spaces. A private outdoor space attached to the master suite will enhance the sense of retreat and relaxation. Whether it's a small balcony, a patio, or a full garden, having direct access to the outdoors can make the master suite feel even more luxurious and connected to nature.

Characteristics of an Outdoor Private Space

- **Outdoor Seating:** A seating area can make the outdoors truly a part of an extended master suite. In a barndominium, the furniture can be made of natural wood and stone to continue the rustic elements in the home. Comfortable chairs, a small table, or even a hammock will provide an intimate spot for morning coffee or an evening view of the stars.
- **Features of Privacy**: The inclusion of privacy screens, fencing, or tall plants would go a long way in making the outdoors private. In barndominiums, as they usually sit on big plots of land, it is mostly private, but one can create an even more intimate setting with trellises or pergolas with climbing vines.
- **Fire Pit or Outdoor Fireplace**: For those who live in cooler climates, the addition of a fire pit or an outdoor fireplace can extend the use of the space throughout the year. This not only adds warmth but creates a focal point that invites one to relax and live outdoors-natural extensions, again, of the barndominium's open, rustic vibe.
- **Hot Tub:** For the ultimate in outdoor luxury, consider adding a hot tub to this master suite outdoor space. Nestled into natural landscaping or placed under a pergola, it can make the area feel just like a spa retreat.

Outdoor Living: Porches, Decks, Outdoor Kitchens.

A place that bridges indoors to outdoors, maintaining the stillness of being outdoors while bringing in the comfort of being indoors. Porches, decks, and outdoor kitchens are some of the many exterior spaces to have gained heightened significance in modern houses; these have especially been found in home designs commonly referred to as barndominiums. The term is a mix of the words barn and condominiums, describing the traditional look of a barn but with modern conveniences to live in. It is not only beautiful to look at but it also functions to give a comfortable place to rest in, enjoy entertainment, or simply enjoy the beauty of nature around it.

Patios: Friendly and Tranquil Areas

The porch is usually the first part of a house that both visitors and residents see. With most barndominiums located in rural or sceneries, porches are the transition greeting from outside the property to inside the home. They are also a sanctuary where one might want to relax and enjoy the morning coffee or simply sit and watch the sunset at the end of the day.

Constructing a porch for a barndominium requires one to seriously consider the functional and aesthetic elements of the porch. Porches for barndominiums are mostly large, running along the front of the house, sometimes extending down the sides to add to the rustic country charm of the structure. Large covered porches in most homes let the owners enjoy the outdoors regardless of the weather outside - they provide a cool shade during hot summer days and protect from rains.

Porches can be as simple or as elaborate as they are, depending on the needs and fancies of the homeowners. For some people, it becomes a real living room as they opt to add rocking chairs, porch swings, or even dining places. Other people may prefer to install other decorative features in their homes such as railings, plants, and lanterns-all these qualities add to the beauty of the residential zone.

Decks: How They Reinvent Outdoor Living

While porches provide a warm and welcoming space at the front door of a house, decks are typically larger platforms for activities that go on outdoors. Most of the barndominiums have decks attached to the back of the house. The decks offer a private retreat with most of them offering a view of the landscape that is all around the house. The actual layout and size of the deck can be very different from each other, depending upon the homeowner's use of the deck.

Decks can be of various natures, with the purposes that they could serve including dining, lounging, and also entertaining. Sectionals and chaise lounges are just two examples of outdoor furniture that many homeowners like to install to create a soothing space in which to bask in the sun. Moreover, decks are very apt for event organizing because one can easily manage the seating area for the guests. The place can be used for a longer period after sunset with the addition of some outdoor lighting like string lights or deck lanterns.

Decks will often be built into the plan, either made from composite material or wood. Both these materials are sturdy and can bear a broad range of weathering. In contrast to composite material that does not fade or rot, wooden decks tend to be more labor-intensive in upkeep, while being one of the oldest materials. Personal preference, budget, and the amount of maintenance homeowners are willing to endure, are just some of the things that need to be considered in choosing the right material.

Outdoor Kitchens

*Post Script: W*ith outdoor kitchens, cooking can now be accommodated outdoors.

The development of an outdoor kitchen creates an elevated definition of outdoor living by providing a functional area to entertain and prepare meals outdoors. More and more, installations in outdoor kitchens are being installed, particularly where the climates are mild and year-round use is allowed. This area often features appliances like barbecues, refrigerators, sinks, and even pizza ovens that allow residents to cook their meals without necessarily going back inside.

Several benefits can be accrued by incorporating an outdoor kitchen into the floor plan of a barndominium-style house. First and foremost, entertaining and hosting becomes far easier. In addition to being a social gathering on its own, guests can congregate around the outdoor kitchen while the food is prepared throughout an event. By reducing the hours that appliances inside the house are in use, outdoor kitchens also keep the house cooler during the hottest summer months.

Outdoor kitchens can be as simple or as elaborate as desired by the user. While some people would be happy with a basic design, such as one that includes a grill and a preparation counter, other people would put full-scale kitchens outside, complete with cabinets, countertops, and even high-end appliances.

Incorporating outdoor living spaces into the floor plan of a barndominium

Floor plans of a barndominium can be adapted in such a way that porches, decks, and outdoor kitchens flow effortlessly into the overall design of the building. Since most barndominiums are designed with an open living concept and in general, are quite spacious, these outdoor areas often automatically extend the home, affording extra square footage for entertaining, relaxing, and enjoyment of the setting.

When designing a barndominium with outdoor living, it should also take into consideration the flow from inside to out. With large sliding glass doors or French doors, they can make for seamless transitions from inside to out of the structure. It might just be what gives someone the feeling of outdoor areas continuing a living room or kitchen.

Whether one uses it to lounge around lazily on the porch, entertain dinner guests on the deck, or cook up a storm in the outdoor kitchen, these outdoor living areas complete the feeling of home in barndominiums.

CHAPTER SIX

Designing for Efficiency and Comfort

When designing the floors of a barndominium, it's all about efficiency with space and cost-cutting measures. Open-concept plans are usually the jumping-off point for every design, which allows flexible living areas with as few hallways and dividers as possible. This is because builders can reduce extra plumbing, wiring, heating, ventilation, and air conditioning systems when they set the main spaces of kitchens, living rooms, and bathrooms as close to one another as possible. This way, material and labor costs are kept minimal. Large windows, high ceilings, and the lack of superfluous details and lines further amplify this light and air circulation, ensuring comfort in an energy-efficient space.

The most crucial thing that an efficient barndominium does, is comfort plays an equal role in such a house design. Insulated walls and energy-efficient windows are options to be installed in homes for indoor temperatures. Beds may be strategically positioned in noisy living areas to serve as peaceful havens. For instance, while increasing the amount of usable space, there is a connection to the natural landscape that encompasses all around by incorporating covered porches and patios into outdoor living spaces. It is done through material selection, like warm woods and neutral tones, that add to the sense of warmth or individual comfort but will also promote aesthetic appeal while balancing it with functionality.

Natural Light: Large Windows, Skylights

Natural light can make or break the look and feel of any room. It influences moods, energy levels, and productivity rates. When the house has enough sunlight, it tends to feel more inviting, spacious, and uplifting. On the other hand, dimly and poorly lit homes may give the impression of smallness, claustrophobia, and even gloominess. Therefore, natural light should not surprise modern architecture, especially in homes with open-concept designs such as barndominums.

Incorporating natural light is yet another feasible alternative, and at the same time, it is also environmentally friendly. You can reduce the number of artificial lights required during the day with skylights and larger windows, thus reducing your month-to-month energy bills and overall environmental footprint. Besides the obvious aesthetic advantages, large amounts of sunlight exposure help regulate the circadian cycles of those residing indoors, which aids in better sleeping, concentration, and overall well-being.

A. Large windows: An added advantage of views

One of the most effective ways to allow plenty of natural light into a barndominium is to design the structure with large windows. In this case, when floor plans are constructed to emphasize high ceilings and open spaces, wide windows will further enhance this openness and the sensation of being connected outside.

- **Layout and it's Positioning**

These many barndominiums offer great views of the countryside, forests, or whatever other natural formations are located nearby. Smart positioning allows the oversized windows to permit the homeowners to enjoy the view and add more brightness inside. Large picture windows may be set in living areas and provide focal points, with easy passage between indoors and outdoors. These commonly would be placed on walls facing spectacular views. Thus, even the most ordinary rooms-the living room, dining, or even the kitchen-become rooms displaying the natural environment.

The orientation of the building could also be another important thing in designing a floor plan for a barndominium; this may, in turn, be able to make the most of the available natural light. Windows facing south would be able to absorb most of the sun during the day, while those facing east can produce a gentle and warm light in the morning. Designers can place windows in areas of the house where they will have the most effect if they pay close attention to how the light passes through the house at every time of day.

- **Styles and Materials**

Barndominiums have a rustic feel to them, yet they are highly contemporary, and many window types can be used to complement this charm. Large windows that extend to the floor are extremely popular, as they create a very dramatic look and allow a great deal of light in. Sliding glass or French doors with large panes of glass are another great option, particularly if the room opens onto a deck or patio.

Multi-pane windows: This may be an excellent choice for those homeowners who would like to incorporate both modern and classic aspects of the home. These types of windows maintain that rustic look generally associated with barndominiums but offer all the functionality and energy efficiency associated with current window technology.

An example of a multi-pane window

- **Energy Efficiency**

Large windows, though aesthetic, may pose a problem in terms of energy efficiency. If not properly insulated, this may lead to a loss of heat during winter and an increase in heat during summer. Fortunately, the technology of modern windows has made great advancements. Several remedies can balance out the negative effects of large windows, including Low-E glass, double panes, and even insulated window frames. While homeowners learn to balance natural light with energy-efficient window treatments, such as insulated curtains or shades, the benefits of natural light don't have to be sacrificed for temperature control in the home.

B. Skylights: Ambient light from above

Skylights are another great way to get natural light inside the barndominium. Where windows will light up the walls and floors of a space, skylights provide the opportunity for sunlight to pour in from above and fill the interior with much more even light. These are particularly helpful for areas where installing regular windows might be a little more challenging, such as hallways, bathrooms, or loft spaces.

- **Placement and Purpose**

Skylights can be very dramatic in barndominiums with high, vaulted ceilings. They allow natural light to pierce deep into the home in areas of the home far from exterior walls. This brightens spaces that may otherwise feel dark or enclosed, like staircases or entryways.

Skylights work wonders in bathrooms and kitchens that require privacy, yet occupants still need a sufficient amount of natural light. In this regard, skylights provide lighting for such areas without giving up privacy. In the case of a bathroom, for instance, with a skylight provided over the shower or tub, the place becomes tranquil and just right for a spa-like experience. Sunlight is allowed inside but at the same time, the privacy of the occupants is still preserved.

- **Skylight Types**

Various skylights are available, each based on the need and style of a home. Fixed skylights are a very basic, economical choice that works for just about any room, allowing natural light inside. Ventilating skylights add fresh air circulation, with their light, which makes them especially great for kitchens and bathrooms where moisture and odors tend to build up.

Tubular skylights are another ingenious creation, handy in those rooms where a full-fledged skylight may not be possible. Compact skylights with a reflective tube bring sunlight from the roof into a room where it's needed. That is great, usually, for illuminating small, dark areas like closets, pantries, or hallways.

- **Energy Considerations**

Like large windows, skylights can impact a home's energy efficiency. However, today's skylights are designed with energy efficiency in mind. Most feature insulated glass and can be installed with shades or blinds to control how much light and heat enters the home.

Blending Form and Function

Adding big windows and skylights into a barndominium floor plan is more than aesthetic; it's a way to create a home that feels bright, open, and connected to nature.

Energy Efficiency: how insulation, energy-efficient windows, and appliances can cut costs

Given that most barndominiums come with open floor plans and big rooms, energy-efficient technologies can be quite handy for such accommodations. At the same time, their unique architecture, being a combination of traditional barn-style buildings and modern-day residences, often turns out to be quite challenging for energy efficiency.

Because these naturally are not insulated metal roofs and large, open interiors- they can be prone to energy loss if their design is not appropriately addressed. It is at this stage that insulation, energy-efficient windows, and energy-efficient appliances come into play.

Insulation: The first line of defense against energy loss

Insulation is the installation process meant to reduce the rate at which heat flows from the inside of your barndominium to the outside. Inadequate insulation makes barndominiums more prone to lose heat in winter and heat gain during summer. This is because most barndominiums encompass big enclosed spaces.

A. **Insulation Forms**
- **Spray Foam Insulation**

Spray foam is one of the most effective forms of insulation in barndominiums. It works very well in sealing gaps and reducing air leaks, which are so common in metal constructions. This insulation type has a high R-value, which is the measure of thermal resistance; hence, it expands to cover fractures and provides a tight seal.

- **Fiberglass insulation**

Fiberglass insulation is a rather traditional type of material; it is easily installed and relatively inexpensive. It can be applied to ceilings, floors, and walls. However, it may not work quite as well as spray foam for sealing small gaps.

- **Rigid Foam Boards Insulation**

Rigid foam insulation possesses a high resistance value. It can be installed in both the walls and roofs of metal buildings. Therefore, the material is particularly useful for metal buildings. This kind of material is extremely useful for barndominiums because they contain wide open spaces.

B. Why Insulation is So Important

Proper insulation can cut down on how long your heating and cooling systems need to run, which in turn will help lower your month-to-month energy costs. For instance, during the summer months, proper insulation keeps cool air trapped inside, thereby minimizing the amount of air conditioning required.

The insulation helps trap heat in winter, thereby cutting down on the need for a lot of energy use for heating. With high ceilings and large, open areas, which characterize a barndominium, reducing heat transmission is crucial concerning ensuring living costs remain lower while still being comfortable. This is because most barndominiums are open areas.

Energy-efficient windows: The permission of natural light to be there while maximizing insulation.

Another highly important feature of energy efficiency is windows. While they give a barndominium ample amounts of natural light, those that are not well-insulated can create tremendous energy loss. Energy-efficient windows will greatly help reduce the cost of heating and cooling a building.

a. **How Energy-Efficient Work**

Energy-efficient windows are generally double- or triple-glazed with more than one layer of glass separated by insulation gas, such as argon or krypton. Such windows hinder heat transfer and maintain indoor temperatures.

Energy-efficient windows also have coatings with low emissivity, sometimes referred to as a low-E coating. This coating reflects infrared radiation inside and allows heat to stay indoors during winter and outside during summer, while still letting in natural light.

b. **Energy-Efficient Windows Being Beneficial to Your Home**
- Energy-efficient windows limit heat loss or gain through the glass of a barndominium that might contain wide windows for views or natural light in open living spaces. In particular, this benefits in improved insulation. The positive impact of this is financial savings in terms of heating and cooling costs.
- Energy-efficient windows reduce drafts and fluctuating temperatures near the windows, which in turn keep your living space more comfortable throughout the year.
- The coatings on energy-efficient windows also block off a significant amount of UV entering your home. UV light can degrade many interior products over time, including flooring, furniture, and other components.

How Energy-Efficient Appliances can Lower Your Electricity Bill

Besides this, another prominent area where energy efficiency may help you save funds or money in your barndominium is in terms of energy-efficient equipment. Appliances like refrigerators, washing machines, and air conditioning units can be major consumers of energy in a household. You can reduce the use of energy without sacrificing performance if you invest in models that are also energy-efficient.

a. **Understanding Energy Star Ratings:**

"ENERGY STAR-rated appliances have met strict energy efficiency standards put in place by the United States Environmental Protection Agency. Energy Star appliances use less energy compared to standard models. This translates into a lower energy bill every month throughout the appliance's lifetime.

- **Refrigerators:** An Energy Star-certified refrigerator runs fifteen percent more efficiently than its similar non-certified counterparts. This appliance could maintain lower temperatures at the set point using less electricity than before.
- **Washing Machines:** Generally, energy-efficient washing machines use less water and electricity than their competition. In addition, they tend to be much gentler with clothes, resulting in less wear and tear over time.

- **HVAC Systems:** Heating, ventilation, and air conditioning systems that are energy-efficient deliver the same degree of comfort while using a significant amount less energy than conventional systems. Speaking of heating and cooling big buildings, this is an especially important consideration for a barndominium because the costs can be fairly high in the absence of appropriate solutions.

b. **Energy Efficiency Impact on Energy Costs**

Energy-efficient appliances can save a significant amount of money. Though the purchase price for some energy-efficient models may be higher, over time, savings in reduced electricity bills can make those investments worthwhile. Additional tax rebates or other incentives for some energy-efficient appliances further lower the cost of the appliance.

Strategies for Incorporating Energy Efficiency in Floor Plans for Barndominium Designs

If integrated during the design stage, incorporating energy-efficient measures into the design of your barndominium can be easier and less costly. Here are some tips on designing an energy-efficient barndominium:

- Arrange windows to allow the maximum amount of natural light while minimizing the amount of direct sunlight they are subjected to during the day's hottest part. This said the barndominium orientation is supposed to capitalize on natural shading and wind patterns where possible to help regulate temperature.
- When you are thinking about heating, ventilation, and air conditioning, you should consider zoning these systems so that energy is not wasted on portions of the house, not in use.

Ventilation and Airflow: how to design a layout for better circulation

Good ventilation or airflow is compulsory to reinforce comfort, preserve health, and increase energy efficiency in any living environment. The layout of a barndominium significantly affects airflow, which in turn impacts several factors for temperature control and the removal of indoor pollutants. One can only optimize living conditions with a good understanding of ventilation and airflow, coupled with appropriate preparation.

Importance of Ventilation/ Airflow

Where there is good ventilation, it is absolute that the air inside becomes stale as fresh air from the outside takes its place. It helps to regulate humidity, reduces allergens, and prevents dangerous indoor pollutants like carbon dioxide, volatile organic compounds, and moisture from building up and fostering mold development. It's such a large structure and all open; being able to control the flow of air or create dead zones is much more important.

Conversely, airflow itself provides a methodology whereby the distribution of temperatures occurs. A good summer circulation can cool spaces through evaporation, while in winter such a circulation will distribute warm air throughout the structure uniformly. Where there is poor airflow, sections of the house may seem stuffy and uncomfortable.

Principles of Airflow and Ventilation in Barndominium or Any Other Buildings

a. Cross Ventilation

This is usually accomplished by putting openings, windows, or doors opposite each other in a room or building to allow natural flow. The wind enters through one side of the building and then pushes the inside air out through the opening on the other side as it continues its course in a continuous flow. Since barndominiums are large, open areas that can take advantage of the wind to circulate air, cross ventilation is one of the best passive cooling techniques that can be applied to such structures.

b. Stack Ventilation or the Stack Effect

This is a means of ventilating by using the natural lifting of heated air and the inflow of cooler air at lower elevations. The stack effect will likely be strong in the barndominium, as they typically have higher ceiling heights compared to conventional homes. You may wish to incorporate vents or windows both low and high within the structure to minimize the number of mechanical cooling needed. This will allow cooler air to enter from the bottom and for hot air to leave from the top, reducing the demand for HVAC.

c. Zoning for Climate Control

In a large condominium that has an open floor plan-such as with a barndominium-same end zoning, separate areas according to the different ventilation or airflow those areas require can go a long way in enhancing energy efficiency along with comfort.

For example, the ability to have more control over the temperature and air quality throughout the house can be accomplished by placing living areas in regions that have the most possibility for cross ventilation and placing beds within climatic zones that are more tightly controlled.

d. Natural Ventilation with Design Elements:

Using certain architectural features can enhance natural ventilation in a barndominium. Clerestory windows installed high on the walls will allow heated air to escape while allowing more natural light into the area. Roof vents will help to release that trapped heat and facilitate airflow.

Practical Instructions for Better Air Circulation in Design

- **Place Windows and Doors**

Strategically one of the primary considerations you can have while designing a floor plan regarding your barndominium is placing the windows and doors such that they face each other directly or at least nearly directly across the space on opposite walls for cross ventilation. Of course, these are handier in effecting the creation of both high and low entry points, thereby maximizing efficiency in the circulation of air. French and sliding glass doors are among those used to open up huge apertures, improving the flow of air between interior spaces and the outdoors.

Smaller windows can be deliberately placed in rooms that require special temperature control, such as bedrooms or bathrooms, to ensure that the inside is well-ventilated but not overly exposed to the outside environment's temperature fluctuations.

- **Use Open Floor Plans for Zero Obstruction to Airflow**

A typical barndominium design already tends toward an open structure, which, by effect, promotes airflow because of no walls and other obstacles blocking the air. This is something already within the design. However, it is of prime importance to organize large things such as storage units, furniture, and other structures so that they will not hinder or constrict air passages. You should apply a careful approach especially not to block any window or door, in addition to making use of light and open shelving or other furniture items that allow air to circulate freely around that area.

Also, it would be advisable to use flexible partitions, such as sliding doors or movable walls, in case an individual needs to be separated yet still have ventilation if necessary.

- **Provide the Ceiling With Ventilation Systems and Fans**

Where natural ventilation works, it is often inadequate to serve the purpose, especially during the hotter months of the year or when the wind hardly blows. As a rule, installing ceiling fans as a basis for inducing air circulation is practical and relevant. In rooms with high ceilings, such as living rooms, dining rooms, or bedrooms, fans should be fitted; this will distribute air well.

Installation of exhaust fans in bathrooms, kitchens, and utility rooms will move moist air out and reduce condensation. A whole-house fan may be installed in the attic or an upper level of the barndominium for pulling hot air out and bringing cool air into the barndominium during summer months.

- **Create Air Pathways through Hallways and Openings**

Barndominiums are typically designed to incorporate long hallways or large open areas that can be used as natural airways. Hallways have to be big and let air pass easily from one portion of the home to another. One can add transom windows or open vents above doors to enhance airflow between rooms. If there is loft space in the design, then the room for air to circulate between the floors remains so stale, and hot air does not build up.

- **Outdoor Ventilation by way of Porches and Overhangs**

Adding covered porches, verandas, and overhangs to your barndominium increases useful space and can contribute toward ventilation. These areas may be considered transitional spaces that allow cooling of the air before reaching the house. Deep overhangs protect windows from direct sunlight, reducing heat gain but encouraging shaded areas and cooler air in homes indoors.

- **Use Breathable Building Materials**

The materials through which your barndominium is built greatly affect the natural ventilation of your home. For instance, natural and permeable materials, such as wood or bricks, have much better air exchange than materials that are tightly sealed. In addition, specific features inherent in most barndominiums, such as metal siding, can be coupled with several insulation and ventilation methods aimed at limiting heat gain and providing circulation.

Barndominium Smart Features: Technology incorporation through smart home features, including convenience features

Today, barndominiums are increasingly integrated with state-of-the-art technology into convenience, safety, energy efficiency, and comfort in general. In fact, technology in lights, smart security systems, and just about everything is changing how home dwellers live and interact with their barndominiums.

This section deeply explores the most common smart features that may be implemented in barndominiums and explains how all this technology improves one's living experience.

A. Smart Lighting Systems

Of the many technologies owners inject into their barndominiums, smart lighting is perhaps the most accessible and prevalent. They permit the owner to turn lights on and off remotely by voice command, through smartphones, or by using tablets. Many times, this is achieved by virtual assistants like Amazon Alexa or Google Home.

Smart bulbs are Wi-Fi enabled; hence you can control them from any location. That means you can turn them on and off when you are not in a space. This feature is mainly beneficial in big barndominiums because managing each light manually can be a hectic task. Moreover, smart lighting systems can be engineered to apply schedules to which they can automatically turn the lights on or off at stipulated times in correspondence with the owner's routine.

Besides the advantages mentioned above, energy efficiency is another added advantage that comes along with smart lighting systems. This ensures the house reduces useless power consumption, leading to an equally reduced electricity bill and reduced impact on the environment. Homeowners can achieve this by dimming the light, setting time, and remote control of lights. Some come installed with motion sensors, ensuring that once there is nobody in the room, the light goes off. This further reduces energy consumption.

B. Smart Climate Control and Thermostats

Maintaining temperature can be tricky in barndominiums, considering the vast, open areas that generally characterize them. The use of smart thermostats presents a solution to this problem by offering precise control over heating and cooling systems. This guarantees comfort in the indoor atmosphere while maximizing energy economy at the same time.

Included in these are smart thermostats that learn automatically the most comfortable temperature settings for a homeowner and modify themselves accordingly. The Nest Learning Thermostat and Ecobee are examples of how these systems can modify the climate at sleep time. They can also be remotely controlled. This means you can control the temperature in your barndominium from anywhere using a mobile app. This comes in handy for those situations where one might be returning home earlier than expected, or when one is staying out later than anticipated.

With smart thermostats, there is also the possibility of receiving real-time energy reports that locate patterns in energy use and help the homeowner make informed decisions to avert energy wastage. Smart gadgets offer financial savings and environmental benefits, which can be quite helpful in a big barndominium where the amount of energy consumed for heating and cooling can be very great.

C. The Intelligent Security Systems.

Most people try to ensure that safety is number one in their homes, and barndominiums are no exception. The usage of smart technologies has surely made security in barndominiums much stronger, customizable, and user-friendly.

Smart security systems usually include motion detectors, alarm systems, smart locks, and Wi-Fi cameras. With these systems, homeowners can view their property in real-time right from their mobile devices, such as smartphones and tablets. With cameras like the Ring Video Doorbell, one can even view live footage, get notifications in case of motion detection, or even talk to visitors. This is regardless of whether you are inside the barndominium or not.

Another key component in security for barndominiums is the addition of smart locks. They finally enable keyless access, whereby doors can be unlocked through voice commands, smartphone applications, and even biometrics such as fingerprinting with these devices. Apart from adding an extra amount of security, it is very convenient for moving groceries or other items within the house. If some of your family members or friends must enter the house when you are not available, temporary access codes can be generated without having to use or distribute physical keys.

D. Automated Shades and Window Treatments

Many of the newest barndominiums feature larger windows for maximizing natural light and views. But there can be an overwhelming task in taking control of the sun and privacy in those large expanses. Automated window treatments and shades can offer a sophisticated solution.

These smart devices can be programmed to operate at specific times of the day or in response to altered weather conditions to avoid disrupting anything. For example, one can mount shades to automatically roll down during the peak heat of the day during summer months when it is hot so that a cool room environment is achieved with minimum air-conditioning. These can be opened when the outside weather is cold to allow sunlight inside and aid in the natural heating of the space.

In addition to comfort and energy use reduction, motorized shades provide a nice level of convenience. Rooms, or even spaces, that could present accessibility problems with higher-than-usual ceilings typical in barndominiums can have their window treatments adjusted with your voice command or by tapping on your smartphone.

E. Smart Appliances and Gadgets for the Kitchen

Since the kitchen has sometimes been considered the heart of any home, it is increasingly seen that barndominiums leverage smart equipment for outfitting. This is convenience and technology to make lives easier in every possible way, like smart refrigerators to tell you when you lack groceries to remote-controlled ovens. Some of the examples involved herein include:

Inside cameras are just one feature of the Samsung Family Hub smart refrigerator, which lets you view the inside contents of your refrigerator from your smartphone while out shopping for groceries. Better yet, they let you create shopping lists, set reminders for expiration dates, and even suggest recipes based on the components that are on hand.

Smart technology equips ovens-such as those from the lineups of GE and Whirlpool-for remote preheating and to signal at the end of a cooking cycle. Such gadgets make meal preparation more efficient, so a homeowner can multitask while managing meals easily.

F. Smart Audio Assistants and Home Automation Hubs

Without the inclusion of a voice assistant or a home automation hub, there is not going to be a fully functional smart barndominium. Devices such as the Amazon Echo and Google Nest Hub are examples of devices serving as command centers for all other smart systems in place. These will, in turn, allow you to run everything from lights to security in your home with your voice or one app.

A home automation hub will connect all the smart devices in the barndominium together for a more smooth living experience. Example routines can be created such that by the time you wake up, the thermostat can adjust the temperature, the lights can gradually brighten up, or a coffee maker can start brewing.

Other advantages of voice assistants include having the devices handled without the use of your hands. These could be when you are in the kitchen preparing food or just relaxing in the living room or even outside the house.

G. Smart Irrigation and Outdoor Features

Because many barndominiums are built onto larger pieces of land, outdoor upkeep might entail a lot of work. For an effective solution, smart irrigation systems use automation that considers weather conditions and soil moisture levels when watering.

These systems are controllable through an application on your smartphone to adjust the frequency of watering or check on water consumption remotely. Some of the characteristics of smart irrigation systems include rain delay and soil moisture sensors. These systems save water and ensure that gardens and lawns are taken care of in the best way possible, even in the absence of their owners.

CHAPTER SEVEN

Choices Regarding Floor Plans

When choosing a floor plan for a barndominium, one must balance functionality with a little personal flair. Consider what you need for your lifestyle: open-concept areas for family gatherings or rooms with clear boundaries for privacy. So versatile are the barndominiums that imaginative floor plans can be designed featuring lofts, home offices, or huge living areas.

Besides that, you would want to consider planning for future extensions or specialized sections, like garages or workshops. In addition, to make your barndominium as functional and comfortable as possible, you have to take into consideration natural lighting, storage options, and ways of going outdoors. Ultimately, the best floor plan for you will be one that reflects both your aesthetic choices and the practical necessities of your living situation.

One-Story Plans: Very affordable and easy to design

One–story plans have easy access and are ideal for a family with small kids or aged members. Also, a single-level home ensures simplicity, accessibility, and affordability qualities combined for an ideal option in many households. Within this section, we deep dive into some of the advantages that occur with one-story barndominium plans. Some of these advantages include the affordability of the plans, the easiness of the construction, and the suitability for families that possess either young or elderly members.

A. **The Level-headedness for Buying a One-Story Barndominium Plan**

Probably, the most relevant argument that speaks to choosing a one-story barndominium is the cost. One of the strongest reasons why people would want to go with a one-story barndominium is affordability. On the whole, one-story homes tend to be more budget-friendly for a lot of reasons when put against multi-story ones. Here are some of those reasons:

- **Lower Construction Costs**: Since one-story houses are usually cheaper than multi-story ones due to several reasons, building a one-story house, you avoid elaborate structural support and complicated staircases, not to mention additional roofing in the case of multi-story options. This ultimately means reduced construction costs. The number of materials used in construction is reduced, and the manpower too; therefore, the cost of the entire project drops drastically. Secondly, a barndominium has an open floor, which can easily reduce the number of inside walls, hence making the costs of constructing even more affordable.
- **Energy Efficiency**: A single-story house is always more energy-efficient compared to a multi-level house since there is a consistent heat distribution around the home. This is in comparison with many other residences with levels. In a one-story barndominium, heating and cooling systems will be able to function more effectively and cohesively, thus allowing for the temperature to be maintained more consistently across the space. Due to such efficiency, utility bills for homeowners go down and thus save them money in the long run.
- **Cost-effective Maintenance**: One-story houses excel in many ways and tend to be cost-effective in terms of maintenance. Maintenance tends to be easier and cheaper since a single story barndominium does not have complications that come alongside many stories. Everything being on one level means fewer areas that require fixing, and any issues relative to the plumbing, electrical, or roofing systems are easier to handle.

B. **Ease in Design and Customization**

Designing a one-story barndominium is easy; it has a lot of options to make it personalized to the needs and tastes of the homeowner without being overly complicated. The simplicity of this layout allows architects and designers to focus their attention on maximizing available space and utility.

- **Open Floor Plans**: Probably one of the distinguishing features of barndominiums compared to all other kinds of homes. The house was designed to have one story, and this open concept helps to free up the flow inside the home and, therefore, make it feel larger and more connected. Inside walls are absent, and movement around the house in different rooms such as the kitchen, dining area, and living room can be very smooth.
- **Flexibility in Room Placement**: Room placement also allows greater flexibility in the way the rooms are set up because everything is on the same level. Bedrooms, baths, and common areas can simply be rearranged for the family's convenience. Due to this flexibility, a homeowner can add larger living spaces, extra bedrooms, or even specialty

sections, such as a home office or playroom, to their floor plan as they see fit for their lifestyle.

- **Personalization**: Perhaps one of the better-known features of a barndominium is its ability to have customized exteriors, and this feature is no exception with single-story designs. Owners can personalize their home by choosing among an array of architectural, material, and finish varieties to make it truly their own. Be it a sleek, modern design, a homey farmhouse comfort, or the classic appeal of the rustic barn style, the options to personalize are endless.

C. **Easily Accessible to Families with Young Children**

With families having young children, the benefit of a one-story barndominium contributes to their level of safety, convenience, and simplicity in a number of the following ways:

- **Safety**: It is one of the most essential concerns for families that have young children, and it gets even more relevant in any situation that involves the staircase. Stairs in a one-level home don't pose any tripping or falling hazard, and thus stairs are one less thing to be worried about. Due to this, there is no need to install baby gates or any other safety measures that would necessarily be used inside a multi-story home.
- **Convenience:** This is much easier for parents to take a glimpse at their kids since all living spaces are on the same level, regardless of where they are inside the house. This is when the parents simply monitor them instead of having to run up and down the stairs, especially when children are playing in the living room, working on their homework in the bedroom, or helping out in the kitchen.
- **Effortless Mobility**: A single-story barndominium makes day-to-day life easy for parents with a newborn or toddler child. This serves to eliminate the need for managing stairs, and by far makes the moving of things such as strollers, infant gear, and even laundry between rooms much easier. Cleaning and organization of a home aren't that much of a nuisance because of this ease in mobility; it will enable parents to focus their attention more on being with their child.

D. **Suitable for the Elderly Members of the Family**

One-story barndominiums are not only suitable for families with young children, but they are also an excellent choice for senior members of the family. They are so comfortable, accessible, and independent when it relates to living situations.

- **Accessibility:** Many individuals in old age face a lot of difficulty in moving around. The elderly can walk around easily without the fear of falling if they are living in single-story homes as they would have no usage of stairs. The house architecture can be made in such a way that it may incorporate wide halls and doorways to accommodate a wheelchair or walker. This will keep the property accessible even later on when the elderly members of the family get older.

- **Independence:** A one-story barndominium offers a sense of independence and is good for aged parents or grandparents who still live with their families. They would be able to get into all parts of the house without requiring any assistance because there would not be any stairs to climb up or go down. An open floor plan also helps in easy navigation, therefore minimizing the chance of any accident or injury.
- **Aging in Place:** More and more homeowners are already thinking of their future by building their homes to age in the place they currently reside. To this end, a one-story barndominium is an ideal option for all the facilities one may need when aging comfortably. Having all of their essential living spaces on one level, homeowners with diminishing mobility may avoid having to make future adaptations or moves.

E. Low-maintenance, Long-Lasting Materials

Barndominiums are made with common building materials such as metal frames and siding. These metals offer several advantages, being stronger and involving lesser maintenance. This is quite useful in those households that look for homes requiring lesser care.

- **Durable**: One-story barndominiums are very durable since siding and roofing are made from metals. From this aspect, the exterior of a barndominium is made from metal, hence capable of withstanding the hazards of extreme weather conditions with very little maintenance. In contrast to this, the exteriors of average residences are made of wood and thus require repainting and adjustments frequently.
- **Resistance to Pests**: Employing metals and other solid materials in the construction of barndominiums avoids the appearance of pests, such as termites and rats, usually present in homes framed with wood. The added layer of protection guarantees the house will stay in good condition long-term, saving the homeowner money and stress in battling problems with pests.
- **Fire Resistance:** Besides that, it further makes the metallic buildings more fire-resistant compared to the traditionally wood-framed homes. This further provides extra protection for families, particularly those families living in states prone to natural disasters in the form of wildfires, among other catastrophes.

Two-Story Plans: Save space by pushing private spaces upstairs

Two-story barndominiums boast several advantages. First, there is the very important ability to save space by using vertically stacked living rooms instead of spreading them out horizontally. This cannot be underscored enough, mainly for houses or lots which may be smaller in size.

For the communal areas such as the kitchen, living, and dining to fit in at the ground level, bedrooms and any other private areas have to be shifted upstairs. Perhaps this separation will allow the house to flow freely; allowing visitors easily or allow open, public spaces on the ground floor while maintaining rooms on the upper floor.

Advantages of Moving Private Spaces on the Upper Ground Floor

This is one of the most marked characteristics: in two-story barn house designs, private areas like bedrooms and bathrooms are usually situated upstairs. Several major advantages go along with this layout:

- **Improved Privacy:** Increased upstairs bedrooms and bathrooms will naturally be much farther away from the rest of the house, making it easier for each member of the family, as well as guests, to maintain their privacy. This design allows homes with children to have a different zone for sleeping and personal activities away from hustle and bustle possibly emanating from the more public ground floor
- **Clear Division between Private and Public Spaces**: By relocating the private rooms onto the upper floor, a distinct public/private divide can be created. This also creates a clear distinction between areas for personal or family use and spaces that are meant to socialize. This spatial structure, especially in those cases where the ground floor is planned to be more open and communal, could facilitate higher levels of noise and activity in the space more easily. Should you have company over, for example, you can entertain them downstairs without the worry that they may be encroaching on your personal space.
- **Floor Space Usage Optimization on the ground floor:** The fact that the bedrooms and bathrooms are upstairs allows the whole ground floor to be relegated for use by shared

living spaces. This allows larger and more open layouts in the kitchen, living room, and dining room. A good example includes how many barndominium floor plans embrace this open-concept trend by using the ground level for vast spaces connected and can house social gatherings down to family get-togethers.

- **More Natural Light:** Besides maximizing the use of natural light within a house, shifting bedrooms upwards allows for better use of natural light. The living areas at ground level could have bigger windows and sliding glass doors while the windows in the upper-floor bedrooms could allow for the supply of light and ventilation onto the beds. Having this kind of design makes a house seemingly open and airy, as more natural light illuminates the space from every angle.
- **More Energy Efficiency:** More importantly, another benefit to moving private areas to the second level can be added energy efficiency. The design of many barndominiums incorporates means for natural heating and cooling. For instance, leaving an open and connected ground floor will naturally increase airflow and may help to regulate the temperature of the space with little need for a great deal of air conditioning. Because heat rises, upstairs bedrooms may stay warm during winter months, and this might also decrease the need for added heating during colder months of the year.

Standard Two-story barndominium Floor Plans House Features

Two-story barndominiums generally put a combination of several key factors together that make them functional and desirable for a wide range of homeowners. The following are some of the most common features one is likely to find in a two-story barndominium floor plan in most cases:

- **Open-concept Living:** Most of the two-story barnominiums have open-concept living on the first floor. Such living is termed open-concept living. Such an architecture eliminates the use of walls and other barriers between spaces such as kitchens, dining areas, and living rooms. Space is more open and allows for increased social interaction. High ceilings, often in the form of vaulted or cathedral-style ceilings, will further enhance the openness and airy feeling common to this type of house.
- **Loft Areas:** Loft areas are among the most common when designing the second floor of most two-story barndominiums. Lofts characteristically overlook the first floor and often double as home offices, extended living spaces, and bedrooms. Loft areas can be utilized as flexible, multi-purpose rooms that may overlook the ground floor area. These rooms utilize the high ceilings common in many barndominium designs, creating a passageway between the two tiers of the house that is both attractive and functional.
- **Porches and Outdoor Living Spaces:** With the construction of barndominiums having an exterior similar to that of a barn, it is quite common to find large porches or wraparound decks as part of the architectural feature of the houses. Such outdoor spaces serve as natural extensions of the ground-level living area for dining, relaxation, and

entertainment, among many other uses. Porches provide residents in more rural settings with a close feeling of connection to nature, further adding to the rural charm of the residence and enhancing its appeal to buyers.

- **Garages or Workshops**: Many of the barndominium plans have additional space for garages or workshops. The garages or the workshops can be adjacent to or even below the main living sections. These rooms would be great for owners who wish to add extra storage or include a workstation for hobbies or even a home company as part of their interior design. This would mean that the living space can be kept separate and noise-free in a two-story barndominium-even when activities other than garage or workshop functions are taking place in those areas.
- **Natural Light and Large Windows:** Large windows to the floor are one of the common features one finds in most barndominium designs, especially within the main living rooms on the ground floor. This allows plenty of natural light to fill in. These windows also act as a means to link the inside spaces and the outside and permit a significant amount of natural light. The bedrooms on the second floor allow light and views making it feel even more that the areas on the second floor are connected with the natural environment.

How to Customize a Two-Story Barndominium

It's an attractive home on many levels but is important for perhaps one reason: a barndominium can fit around your needs concerning lifestyle. You have even more room with a two-story design to decide how you might want to define the layout or distribute space. Of course, many homeowners choose to add additional bedrooms upstairs, while others like to leave the upstairs open and perhaps use loft spaces for whatever purpose.

Another plus is that no interior walls bear weight because the barndominiums are often built from metal building kits. This allows for more flexibility in terms of the changes you can make with interior design and layout.

L-shaped and U-shaped Layouts

This is an ideal plan for the flow between indoors and outdoors; courtyard designs included.

Architecturally, the layouts of houses and buildings play a monumental role in being able to do so in how spaces interact with one another and more importantly how they transition between indoors and outdoors.

The L-shaped and U-shaped layouts are particularly beneficial to promoting a harmonious flow between the inside and exterior of a building. Of the many varied types of layouts that are available, such types of layouts have a particular set of advantages. Not only do such designs realize the full aesthetic and functional potential of courtyards, but they do so in a way that creates spaces that invite nature inside, helping to foster a smooth interaction between the built environment and its natural surroundings.

Knowing the Difference between L-shaped and U-shaped layouts

The difference in the configuration of the L-shaped and U-shaped layout. By nature, these below-mentioned layouts create a semi-enclosed space that is most often arranged in the form of a courtyard. This semi-enclosure will foster stronger communication between the interior living area and exterior spaces it is paired with by its very nature. The striking distinguishing feature between the two is the tendency of both to attract eye and activity towards a central exterior space, usually a patio, garden, or courtyard.

A. L-Shape

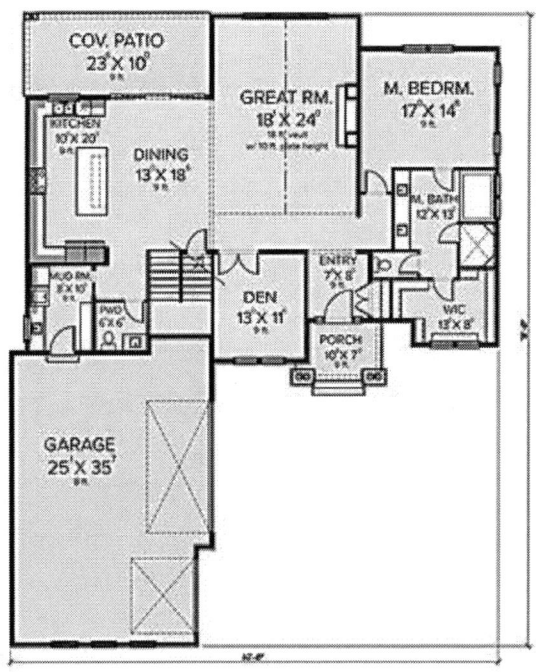

The L-shape design consists of a design wherein both wings of the structure running perpendicular to one another form an "L" shape, thereby creating a partial enclosure. The open side of the "L" can often look like a garden or patio and provides unimpeded views and access from a variety of rooms around the house. This plan works particularly well on smaller sites or homes wanting to maximize the connection to one side of the land.

B. U-Shape

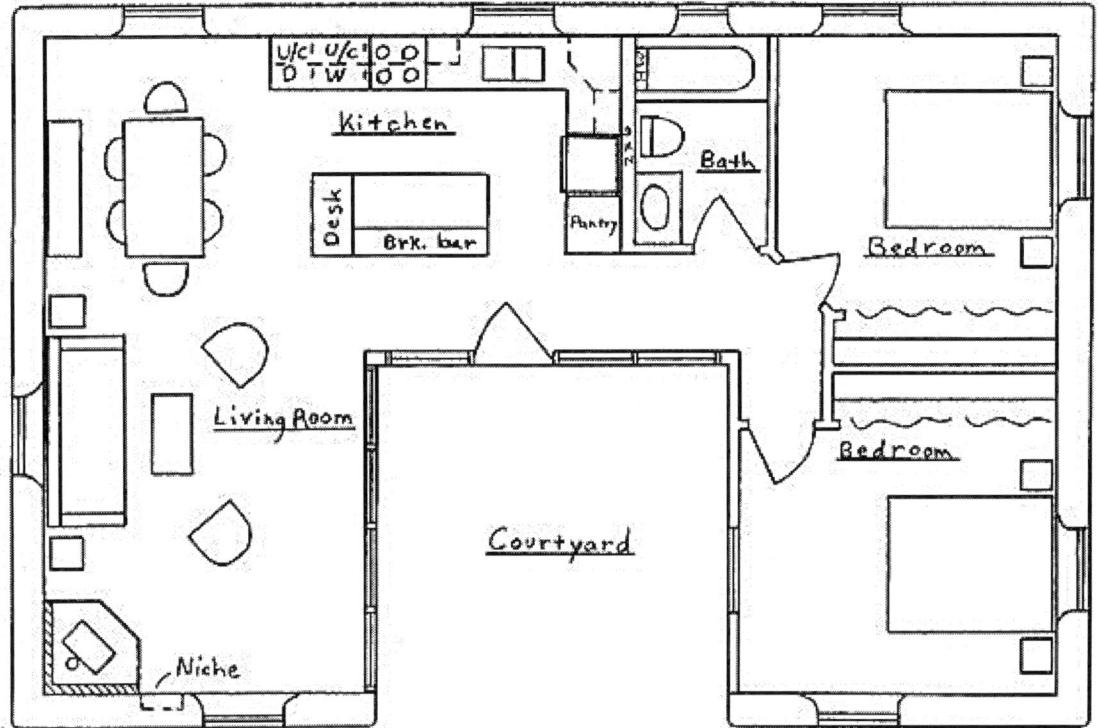

The U-shaped type is wider. It covers three sides, leaving an outdoor area in the middle. This structure is perfect for big houses, wherein it provides a greater degree of seclusion without impeding adequate ventilation of sunlight into interior rooms. Courtyards can be used frequently as focal points in such a design because they provide a sensation of being in solitude yet feeling near the interior space.

Improved Flow between Indoors and Outdoors

One of the great advantages of L-shaped and U-shaped layouts is their flexibility in terms of providing a smooth transition between interior and outdoor spaces. In such cases, placing strategic doors and windows, including open-plan areas, may help achieve a cohesive life experience while dissolving the boundaries inside and outside a building.

- **Visual Continuity**: All floor plans provide for the installation of huge expanses of glazing, such as floor-to-ceiling windows, sliding glass doors, or bi-fold doors, which create a continued view between the indoor living rooms and outdoor areas. This viewed continuity makes the perceived space greater, thereby creating an impression that indoor spaces seem larger and open.

- **Easy Flow:** Both L-shaped and U-shaped homes have a functional flow to allow easy movement around. This is because the outdoor space would be directly accessible from multiple inside rooms. For instance, the kitchen will probably be at one end of the L-shaped house, while the living area is at the other. In turn, both areas open out to a patio shared by all the family. This design is ideal for holding events as it creates an easy and free movement of people between the indoor and outdoor areas.
- **Maximizing Natural light:** These layouts catch the most daylight, as the open side of the L or U faces the sun, in turn spilling inside and minimizing artificial lighting. This helps to integrate the interior much further into the natural world outside, while at the same time contributing to the creation of a pleasant mood inside.
- **Creation of Microclimates:** This courtyard or garden that is created within these design elements acts as a microclimate, making it insensitive to wind and noise effects. In turn, this develops a more controlled outdoor environment, whereby outside comfort for the residents is significantly enhanced without the most injurious elements of the weather. Second, it helps to regulate the temperature by the existence of water features, plants, and trees in courtyards, offering the cooling effect naturally during hot months and offering a warmer microclimate during that time of the year when the temperature is usually low.

The courtyard designs: The heart of L-shaped and U-shaped homes.

The courtyard has been a great feature in architecture since ancient times, from the Roman domus to the Chinese siheyuan dwellings. In modern architecture, it is still an important position and many of them form part of L- and U-shaped layouts, often forming the focal point of the overall design. There are several advantageous features that courtyards offer, supplementing the transition between interior space and outdoors.

- **Solitude and seclusion**: If the courtyard is placed in between the arms of the structure, then the homeowner can have an outdoor space that is kept hidden from the general public's view. With such a setting, the courtyard becomes an ideal location for relaxation, dining, and even extending events to take place outdoors.
- **Natural ventilation**: One of the ways that courtyards allow the improvement of natural ventilation is by creating space that allows easy circulation of air between different parts of the house. Such is the case with U-shaped layouts, for example, where during the warmer months, the air can flow naturally around the structure to cool the interior of the building. In other words, design provides advantages using passive cooling principles, which can help lower the need for air conditioning.
- **Bringing the outdoors life**: Incorporating landscaping, water features, and even art into the courtyard brings the outdoors in as an extension of the indoors. The beauty added to the home by such elements is only the tip of the iceberg in the good effect on the tranquility and well-being of the occupants. Besides, courtyards can also be designed to lure local animals like birds and butterflies which would help further their relationship with nature.
- **Year-round usability:** The use of courtyards can be extended for all seasons if designed thoughtfully. For example, the incorporation of retractable awnings, pergolas, or space heaters in the area will make it comfortable during not-so-wonderful weather. The fact that these houses can be used all year round makes L-shaped and U-shaped houses very flexible, in that outside spaces are not limited to specific times of the year.

Making the most of the views and accessing nature

Framing of views, which is the most important factor for consideration in house design, intended to provide continuity between interior and exterior space, is in L-shaped and U-shaped floor plans. The homely L-shaped can be designed to take advantage of beautiful views by orientating the open side to take advantage of the garden, pool, or landscape beyond the property. Even more than the traditional layout, the U-shaped design sets up a private sanctuary. In fact, it is the central courtyard that takes center stage for nearly all the rooms of the house.

Besides, these layouts are ideally suited for the implementation of biophilic design concepts that emphasize incorporating natural elements into the built environment. These plans develop a deeper contact between the occupants and nature by filling indoor spaces with greenery. This can be achieved by either using courtyards, green walls, or indoor planters, which could help foster both physical and emotional well-being.

Sustainability-related issues

L-shaped and U-shaped configurations facilitate a smooth aesthetic and functional transition between interior and exterior spaces but also enable environmentally responsible design. Because these layouts create a partial enclosure, it becomes significantly easier to apply passive solar design methods. Passive solar techniques use the independent natural heat and light of the sun to reduce the amount of energy used. The maximum winter solar gain with the protection of the interior from summer overheating can be obtained by homeowners located in the northern hemisphere with a simple orientation of the open side to the south.

Also, with the involvement of a courtyard or some other outside area, it makes it easier to add rainwater collection devices or even solar panels, whereby again the impact of size on the home to the environment is diminished. The courtyards inherently offer a cooling effect, and better ventilation possible in L- and U-shaped layouts help reduce artificial cooling systems, thus increasing energy efficiency generally.

CHAPTER EIGHT

Practical Spaces in Barndominiums

With more open-concept layouts, flexible floor plans in a barndominium can make for functional spaces with combined uses and contemporary living. Most houses in barndominium have ample living areas that are smoothly integrated with the kitchen; hence, one can use it for varied purposes.

Another very helpful feature of the barndominiums is the amount of storage provided with either an attached barn or a garage, which is great for those people who possess a lot of equipment or need a workshop area. The living areas, including bedrooms and toilets, can be adjusted accordingly with bathroom or office facilities to meet the needs of the dwellers. In short, this living space is highly adaptable. Besides giving more opportunities for both domestic and industrial use in one place, the strong metallic structure ensures that it will serve much longer.

Garages and Workshops: Attached or detached

Originating first in the rural regions, barndominiums have outgrown their agricultural inception and found favor in suburban and urban areas, as well. It is a prevalent convention to build such houses on a metallic framework that offers several advantages regarding affordability, flexibility, and strength.

With the very high level of customization possible with barndominiums, residents can also create versatile rooms suited to their lifestyles. It is not uncommon for them to have big garages or workshops for the storage of automobiles, workspaces for hobby pursuits, or even commercial enterprise operations. This versatility will be a major selling point to those homeowners who wish to add functional conveniences to their living areas.

The Attached Garages and Workshops

A. Convenience and Proximity to the Location:

Perhaps the most important advantage associated with having an attached garage or workshop to your house is that of accessibility. If the garage or workshop is attached to the main residence, then the homeowners have the convenience of entering the space without having to go outside.

Of course, this is a definite plus, especially in some areas of the world where weather conditions may be quite severe, where going outdoors and dealing with rain, snow, or extremely high temperatures may not be desirable at all. Speaking of the ease of having an attached garage, whether you be gathering tools for a project or parking your car after a hard day, they are irreplaceable and helpful.

When it comes to workshops, having a building that is attached to your home may facilitate easy movement between your working space and your home. This could be an ideal solution in instances where there is a need for frequent visits between the living area and the workshop for individuals who have hobbies or run small enterprises from their property.

B. Cost Optimization:

Sometimes, building an attached garage or workshop is less expensive than creating a detached one. Since at least one wall is shared with the house, the materials and labor involved are already reduced. In addition, you might be able to share HVAC, plumbing, and electrical systems, which will further reduce overall costs.

C. Additional Safety and Security:

One of the other apparent benefits of having a garage or workshop attached to your home is the added security one will derive from such a building. Perhaps, because it is attached to the house, it will be easier to keep tabs on and secure.

This added level of security might provide a homeowner with peace of mind when they are storing valuable objects or vehicles in their garage. In addition, many attached garages provide homeowners with direct access to the home. In this way, residents can avoid the need to traverse across a dark yard when it happens to be nighttime or inclement weather.

D. Easy Aesthetic:

From a design standpoint, the garage or workshop might have an integrated look because it's part of the general structure. For those individuals who want a smoother look, an attached option affords them the possibility of a consistent design sympathetic to the architectural style and material type used in the house.

On the other hand, attached garages and workshops have a fair share of drawbacks that are worthy of consideration: major concerns would include noise. Chances are, proximity to the workplace where loud machinery like saws or drills might be in use may become problematic.

This can be minimized by soundproofing, but it will again increase the overall cost. An added concern, if proper ventilation is not provided, fumes and odors that stem from cars, chemicals, or other objects that have been used in the workshop sometimes seep into the house.

The Detached Garages and Workshops

- **Separation of Spaces**

In such a case, where the garage or workshop is detached from your house, there is a line of demarcation between your living area and your office. Such could be a better situation when the owners of the house like to separate their professional life from their personal life. For example, if your workshop were to be used for noisy or particularly messy activities, then situating it detached ensures that these aspects do not become an annoyance in those parts used for living.

The separation of workshop space from the rest of the house is of utmost importance to those operating businesses from home or who professionally utilize their workshop. This is because the ability of customers or employees to access the workshop without having to make entry through the primary residence causes increased privacy and professionalism.

- **Increased Adaptability with Increased Usage of Space**

Detached garages and workshops are more flexible for land use. This would enable landowners to optimize lot usage to their liking, as these could be located anywhere on a lot. An example would be that you can build a larger detached workshop, one that's further away from the home, without having to make some sort of compromise in the aesthetic harmony of the structural limits.

Another fantastic thing about a detached building is that its extension is easier compared to the attached. Since it is not attached directly to the main house, you will have the leeway to make changes in its size, structure, or style without influencing the general architecture of the house.

- **It Increases Personal Privacy and Security**

On the other hand, a detached structure can be a little safer for homeowners who frequently use power tools or store hazardous products in their workshops. This can dramatically reduce the chances of any accidents or hazardous gas exposure inside the house by isolating such activities. Moreover, keeping the workshop away from the main house can save it from damage in case of a fire or any other kind of calamity happening in the workshop.

- **Noise Control**

A detached workshop provides the best avenue to minimize the level of noise pollution in your house if your work involves loud machinery or equipment. Your work will not disturb the members of your family or the others who stay with you in the same house, and you will be able to concentrate on projects without concerns about creating annoying noises. Again, the distance between home and workplace could offer a psychological buffer that enables you to switch between working and resting mode more efficiently.

- **Impact on Eyes**

One of the major cons that detached garages or workshops will do to your property is to impede the line of vision. These buildings can make a property seem fragmented or disjointed, largely depending on their setting. If the landscape and design elements are carefully thought out, however, a detached building can fit into the property much more graciously.

- **More Expenses**

The initial expenses of construction will possibly be more considerable when dealing with a detached structure. This is mainly because of the higher needs concerning materials, labor, and individual utility hookups. Unlike an attached garage, the detached workshop would require running individual electrical wires, plumbing, and heating systems. With these added costs, the detached option might not be quite as appealing to a homeowner with a modest budget.

Attached or detached-when considering a barndominium floor plan boils down to personal needs and preference of the homeowner. In this case, if you value ease, economy, and continuity of design, an attached may be the better choice. If you are looking for certain things in your life such as privacy, flexibility, or a clear distinction between your professional life and your home life, then a detached structure would more suit your way of living.

Ultimately, your decision should be based on how you would want to use that space, what the weather conditions are around you, and the property layout. What's great about barndominiums is the ability to create flexible spaces. That being said, let's now talk about how to design a garage or workshop to reinforce both your living experience and also the practical needs you have.

Storage: Mudrooms, Laundry Rooms, Attic Storage, and Basements

Storage facilities within the floor plans are thus a necessity in ensuring that maximum space is used for added functionality. These areas, including mudrooms, laundry rooms, attic storage, and basements, all play an important role in the successful management of a clutter-free and efficient house. This essay, therefore, will discuss the type and nature of different storage spaces that most barndominium floor plans make a point to include, their importance, and how they can be best designed and used.

The Mudroom: The Transitional Space

A mudroom is usually a simple space located near a home entrance where one drops off their shoes, jackets, and bags. A mudroom inside the barndominium is necessary due to its function as a transitional site between the area where one has to live and the outside setting, thus keeping the space in the main areas tidy.

- **Design and Functionality**

A mudroom's number one job is to keep dirt, muck, and various other detritus out of the house. Flooring is often vinyl or tile, for its durability and ease of cleaning. It's not strange to find hat and coat hooks, cubbies for shoes, or benches that also double as storage units. When it comes to barndominiums, most of them are constructed in rural locations or on big plots of land, thus accommodating mudrooms, which can be especially useful for families that have dogs, farm equipment, or outdoor gear needing a specific location for storage.

- **Transforming Storage in a Mudroom**

A mudroom has to be created with storage that fits the specific needs of the house. Families with small children may need additional bins for toys, school items, or sports equipment. Moreover, those that would remain on farms might require additional shelves for gardening materials, boots, and other tools used. The flexibility of barndominium open floor design makes it possible to make larger mudrooms. This allows the installation of custom cabinets or even small closets that may be put to use in storing cleaning supplies, extra gear, or seasonal products.

Laundry Rooms: Multi-purpose Utility spaces

Laundries are multifunctional utility rooms serving more purposes than just washing and drying clothes. In barndominiums, laundry rooms often serve many other purposes than what their name tells. In today's modern floor plans, a laundry can also be used as an additional mudroom, a place for dog care, or even a mini-workshop in homes where such needs arise.

- **Plan a Perfectly Designed Laundry Room**

A well-organized laundry room not only makes it easier to perform household chores but also tends to keep the rest of the house clutter-free. It would include ample counter space for sorting and folding clothes, cupboards or storage for detergents and cleaning supplies, and space for hampers and baskets.

The laundry rooms in barndominiums can be larger than in a regular residential abode. Additional storage could be possible with this size: wall-mounted drying racks for clothes, cabinets to store cleaning supplies, and even an ironing board built right into the structure specifically.

- **Expand the Laundry Room Storage**

This laundry room could even serve the owner of a barndominium as a hub for objects that don't have a designated place anywhere else in the house. Things like seasonal clothing, extra bedding, or bulk cleaning items might find a home in the laundry room cabinets or closets. Adding shelves above the washer and dryer is a simplified way to add space, and pull-out baskets or hampers further simplify sorting. To ensure that every inch is used to maximum capacity, custom-made and designed storage containers to fit the unique dimensions of the laundry room are put to work.

Attic Storage: Using the Space Overhead

The attic is one of the most underused spaces in any home. However, in a barndominium, it can be converted into convenient and practical storage. Areas of the attic can be accessed by using a pull-down ladder or even a staircase. This is a very good place to store seasonal items, holiday decorations, or household items that one does not frequently use.

- **Important Design Considerations for Storage in the Attic**

Designing the storage in the Attic and accessing it with ease may mean already successful attic storage. Given this, designs from careful consideration are often necessary, that is, designs for proper ventilation are usually considered to keep objects that will be stored away from extreme temperatures and moisture.

The lighting is also quite important; it's hard to find objects in a completely dark attic. With the higher ceilings and often larger overall structure of barndominiums, the attic can be designed for use other than storage. Others use a portion of the attic space and convert this into an extended living area, home office, or even a guest bedroom while leaving the remaining space for storage.

- **Organizing the Attic**

Other things that go the extra mile in keeping it organized are attic shelving, bins, and labeling. To be able to store heavy boxes, solid shelving can be installed into the frame of the attic, while clear plastic bins offer an easy way to see what is inside without having to rummage through everything. Creating zones for different types of items-holiday decorations, old clothes, keepsakes, or furniture-make it easier to keep track of what is stored in the attic.

Basements: Expansive Storage and More

Traditional barndominiums are usually on slabs and don't have basements, but when they do fall within the design of a barn home, they offer astonishing storage potential. Not only does a basement increase the amount of living space, but it also allows for items that are large or little used to be stored since they cannot fit elsewhere in the home. Whether finished or not, basements may house anything from extra appliances and home gym equipment to seasonal sports gear and camping supplies.

- **Benefits of Basement Storage**

Aside from giving your home some breathing room by storing items in the basement, a well-planned basement is one sure way to keep clutter at bay. Items like off-season clothes, holiday decorations, and furniture are ideal to be stored in the basement. In addition, basements in barndominiums can become multi-service areas: home office space, workshop area, or even a game room while yet offering ample storage space.

- **Maximizing Basement Storage**

In an unfinished basement, shelving units, hooks, and pegboards can make it functional at once; for finished basements, custom cabinetry or closets do the job seamlessly and unobtrusively. It is essential to fit them with dehumidifiers or ensure proper insulation to save the stored items from the awfully common problems in the basement environment and mold.

Vehicle and RV Storage: Room for Cars and Recreational Vehicles

The ownership of more than just a car has grown increasingly common as the population's interests and activities grow in variety and number. Motorhomes, boats, trailers, and off-road vehicles fill up driveways of homes all over the world. Not everyone, however, has the space to house such large items in their yard or property. A vehicle and recreational vehicle storage facility provides a very convenient yet effective solution for outdoor lovers who cannot find space for their cars.

Emerging demand for automobile and recreational vehicle storage

During the past decades, owners of RVs, boats, and other large-sized vehicles have emerged quite dramatically. The RVIA thinks that the number of U.S. households owning recreational vehicles is at an all-time high, with millions owning RVs. This can be attributed to several reasons: an increased popularity of outdoor activities, extension of remote working that allows for greater mobility, van life, and minimalistic living.

As ownership increases, so does the more complicated problem of storage. While most residential properties are not in a position to handle big vehicles such as recreational vehicles or boats, often the local zoning rules or even the HOA regulations may forbid parking huge cars in driveways or yards, even though the particular person may have space enough. These vehicles must be left on the street, which is illegal or too unsafe due to the risk of theft, vandalism, or an accident involving the vehicle. Vehicle and recreational vehicle self-storage facilities become a logical solution in such cases.

Different types of automobile and recreational vehicle storage options

The situation will warrant either the type of vehicle or the level of security that is required for storage solutions. Owners can make an informed decision based on their needs and budget by considering these options.

1. Outdoor Storage

The most common and affordable methods of auto and recreational vehicle storage are outdoors. Generally, this type of storage involves an open lot where autos are positioned according to designated locations. This type of storage offers the least protection against weather conditions, but it allows ample room for the largest-sized recreational vehicles and boats. They face the sun, rain, snow, and hail phenomena that could make them deteriorate little by little.

Outdoor storage would be a good alternative for those on a tighter budget or when the owner uses the vehicle frequently enough not to be concerned about minor wear and tear. Because it does not involve extreme weather conditions, places with mild climates have taken to this method most favorably. Tarps and covers on vehicles or RVs can also serve as protection against the exterior weather elements and decrease the level of damage there is.

2. Covered Storage

The covered storage will allow compromise for those people seeking cost-effectiveness and still offer added security for their automobiles that they would want to have. Vehicles are parked in covered storage, which consists of a structure with the resemblance of a carport and has a roof over it. Such storage minimizes weathering, such as fading, cracking, or rust, since it is shielded from direct sunshine and precipitation.

It is more expensive compared to outdoor storage, but it does have a considerable advantage in extending the life of a vehicle's paint, tires, and interior material. This is the case because covered storage requires less upkeep. Individuals who just use their recreational vehicles seasonally or on an occasional basis are those who would want to make sure their investment stays in good shape between excursions, so this type of storage works well with these individuals.

3. Enclosed Storage

The most protected and out-of-the-elements vehicle and RV storage is indoors. Vehicles are kept in an entirely enclosed facility and often are enclosed in individual units. A facility like this would not only protect against environmental elements but also possible theft, vandalism, and pests. In large part, indoor storage facilities are climate-controlled and add further protection to the vehicle from being damaged due to extreme temperatures, humidity, and moisture.

Indoor storage is the most expensive, but it gives owners an added sense of security in knowing their vehicle's best protection. This is particularly the case for those owners who have invested a great deal in their classic cars or RVs. Indoor storage is normally used for luxurious cars, historic cars, and RVs that are stored for quite a long time, particularly during the off-season.

4. Custom RV Storage Facilities

Specific examples are specialized storage facilities that can accommodate recreational vehicles and large-sized automobiles. For these facilities, one would almost expect to find lots of wide drives, electrical hookups, dump stations, and even maintenance services. Access to this infrastructure is quite important for the owners of the RVs given that they may well need it to undertake minor repairs, recharge their batteries, or drain their tanks in preparation for their next trip.

These will come in very handy for those living in cities where space is an expensive commodity or people staying in regions that tightly restrict the storage of RVs at home. Some high-scale facilities also include concierge services, where getting the recreational vehicle ready for use, cleaning, and filling with fuel are included. This makes it stress-free for the owners of the RV.

Advantages of Using Vehicle and RV Storage

A. Protection from Bad Weather

The reason most people store their vehicles and recreational vehicles is because this provides much-needed shelter from bad weather. Vehicles that have been subjected to bad weather would eventually incur significant damage after some time. Paint will fade, dashboards crack, and upholstery can be ruined by the sun's ultraviolet rays.

Rain and snow can bring accumulation that will cause rust and water damage. Owners can make their automobiles last a bit longer and avoid costly repairs if they can keep their vehicles in settings that offer proper protection.

B. Security and Safety

One of the many wonderful benefits of using a storage facility is its security. Many vehicle and RV storage facilities have video cameras, gated access, and on-site workers that will reduce the possibility of theft or vandalism considerably.

Often, indoor storage containers will feature a lock on each unit, allowing an even greater level of security. This allows the owners to go and leave without their recreational vehicles, knowing their car is in safe and secure territory.

C. Compliance with the Locality's Regulations

Many communities, along with municipalities have developed strict regulations in regards to the storage of recreation vehicles-or RVs, boats, and trailers on residential properties. Generally, a homeowners' group, sometimes referred to as an HOA does not allow large vehicles to be parked in driveways or on the street for extended lengths of time.

Automobiles stored at a facility designed specifically for that purpose ensure compliance with local rules and avoid disagreeable situations with neighbors.

D. Creating Home Space

Self-storage for vehicles makes much more sense when it comes to freeing up home space, especially when the available space in the driveway or garage is at a minimum, which can't be overemphasized, particularly when you have homeowners owning more than one car or live on small lots.

Storing an RV or an extra car could be of help in any facility to create more rooms for cars employed daily, outdoor equipment, or even changes that are made to the homes, such as a garden or a playground.

E. Easy Accessibility

In fact, many storage facilities offer access at any time of the day to enable owners to retrieve their automobiles at whatever time they may need. This flexibility is essential to some people who use their RVs or boats quite frequently and do not want to bear the hassle of towing them from their homes or the hassle of parking on the street.

Things to Consider When Choosing a Storage Facility

Several variables need to be considered when selecting a storage facility for a vehicle or RV to make sure the storage solution will work for specific needs.

- **Location**: A location closer to home, or on routes traveled frequently, may make access more convenient to the vehicle when needed.
- **Price:** The cost of storage would depend on the kind of storage, whether it is covered, indoor, or outdoor; and it would also depend on the amenities it has to offer. There should be an inter-play in balancing the extent of security desired vis-à-vis with the cost.
- **Security:** Be cautious when settling for facilities with so much physical security such as entrance gates, cameras, and employees on location.
- **Amenities:** Some facilities offer additional amenities, such as RV servicing, wash stations, and electrical hookups, that can make a big difference in a motorcycle or other recreational vehicle owner's final decision.
- **Climate Control**: For those with extremely valuable recreational vehicles or classic cars, a climate-controlled inside unit may provide that extra layer of protection against extreme temperature and humidity fluctuations.

CHAPTER NINE

Accessibility Considerations

In developing a floor plan for a Barndominium with accessibility concerns, one needs first to make the facility friendly for those who cannot move around with ease and for those with some other disabilities. These include the installation of larger doors and passageways for wheelchair passage, single-story plans or a living room without stairs, and openness in the floor plan to enable mobility.

Counters in kitchens and bathrooms should be lower with accessible sinks and roll-in showers installed. Additionally, door levers and smart home technologies can make the space easier to navigate. With thoughtful consideration of these variables, one can ensure functionality in space and make it inviting for people of different backgrounds and identities.

Aging in place: Designed to be Comfortable and Mobilized for Extended Periods.

The concept of "aging in place" is a popular phenomenon whereby senior citizens can continue to live in their residential homes for as long as possible and preserve their independence. Because people's mobility and physical needs change with aging, it is more important than ever that their living spaces be modified to accommodate their progressing needs. The barndominium has grown in stature to become an endearing option for many, who look forward to a house that is as comfortable as it is accessible. A barndominium is a generic precast structure fusing the functionality of a barn with the convenience of a house. A barndominium offers huge space and several opportunities for personal touches.

When one is designing a floor plan for a barndominium intended mainly for aging in place, one has to turn its mind to creating a space that caters firstly to safety, effortless mobility, and a livable environment in the long term. A barndominium can be an excellent choice for senior citizens who expect to stay indoors and enjoy the openness, and flexibility offered by the premises while still being able to provide comfort and accessibility as required for extended periods. This is possible with proper planning.

- **Single-Level Housing: Prioritizing Accessibility**

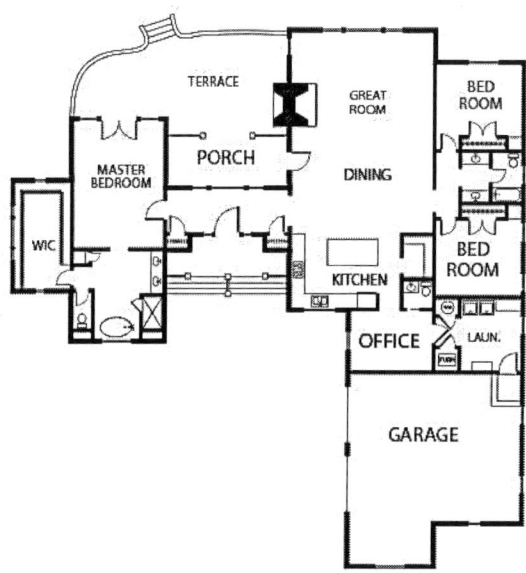

A single-level floor plan is one of the most essential ingredients for a home designed with aging in place. This is because when the mobility of an older individual becomes compromised, the biggest problem faced by many is the use of stairs. Such construction is safer and, at the same time, more comfortable to live in for a person who has some difficulties with his or her mobility, as it means a person will not climb any steps to live in a one-level barndominium. This is the preventative measure that may be taken by people who are still able to move somehow to live safely without some potential dangers appearing shortly.

Since the floor plans of barndominiums are so expansive and open, they make for an ideal selection when it comes to single-level living. Wide-open floor plans are also aesthetically lovely, as they serve to create a space that allows individuals to better maneuver around and also provide ease for allowing in mobility devices such as wheelchairs and walkers. Hallways and entrances should be wide enough to consider safe passage-always 36 inches or wider-and open spaces are to be free of obstructions except for whatever is necessary.

- **The Open-Concept Areas: It Offers Advantages of Flow and Visibility**

As indicated above, an open-concept design should also be one of the important elements of a barndominium floor plan that allows aging in place. Among the key benefits of open-concept plans is that they eliminate superfluous walls and barriers between rooms, which in turn provide much greater visibility across the whole house. Such improvements make it easier for persons in their older age to pass from area to area without incident, whether they are walking on their own, making use of a cane, or dependent upon a wheelchair.

Besides increasing the level of mobility, open-concept spaces create a living space that is more social and interactive for community members. The kitchen, dining room, and living room can all flow into each other, freely allowing family members or caretakers to interact without hindrance. This is hugely important for those who may need care and support in their own homes, as it enables the caregiver to be nearby without intruding.

- **Fully Accessible Kitchens: Practicality and Safety**

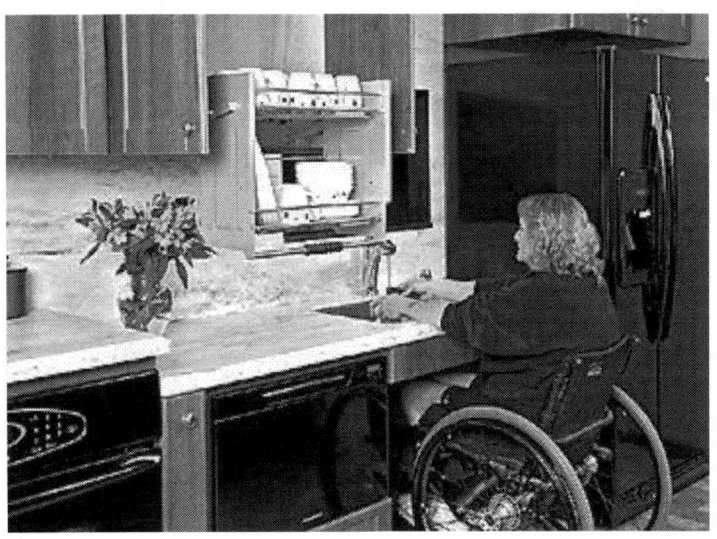

As the kitchen is typically referred to as the "heart" of the home, this room is especially crucial to be functional and safe for seniors aging in place. A barndominium designed for aging in place should have a kitchen that offers easy access to appliances, counters, and cupboards with a minimum of bending, reaching, or lifting.

Given the incorporation of features such as pull-out shelving, counters with varying heights, and easily accessible storage, for example. The ability to comfortably cook with your seat or standing is further facilitated by the incorporation of kitchen islands or counters that are both lowered and variable in height. Safety and ease of use in the kitchen can also be enhanced with better mounting of appliances: wall ovens and side-by-side refrigerators, placing induction cooktops-which reduces the chance of burns.

Another feature is the ease of access to cabinet doors and faucets with lever-style handles. The handles allow persons affected with arthritis or those with minimal hand strength to operate them. Under-cabinet lighting and task lighting are types of lighting that can eliminate shadows and ensure that work areas are well-lit. Lighting is also a vital element when designing a kitchen.

- **Universal Design Features for Bathrooms: It Makes the Need to Prioritize Safety and Independence Paramount**

Bathroom spaces are considered among the most significant areas to focus on when constructing a barndominium with the intent of aging in place. At some point, standard bathrooms may be an accident waiting to happen for an aging senior due to slippery surfaces and cramped quarters.

Bathrooms designed to accommodate individuals aging in place should incorporate universal design principles that emphasize accessibility and safety. Showers should not have any thresholds because one does not have to step over the lip or ledge of a shower threshold, thus reducing the chances of a fall. Grab bars can be installed in the shower and around the toilet for additional support. This can be done in a practical and/or an aesthetic manner.

Those with diverging needs regarding mobility will enjoy a curbless walk-in shower with a bench seat and handheld showerhead for increased accessibility. The floors should be non-slippery, such as textured tile or even rubberized surfaces, to reduce the chances of falls resulting from slipping. Toilets should be installed at heights comfortable for users. Bathroom vanities should be constructed in ways that make their use easily accessible. Moreover, there should be knee space available under the sink for people who would like to use the sink while seated.

- **Wide Doorways and Hallways: It Makes Movement within the Building Easy**

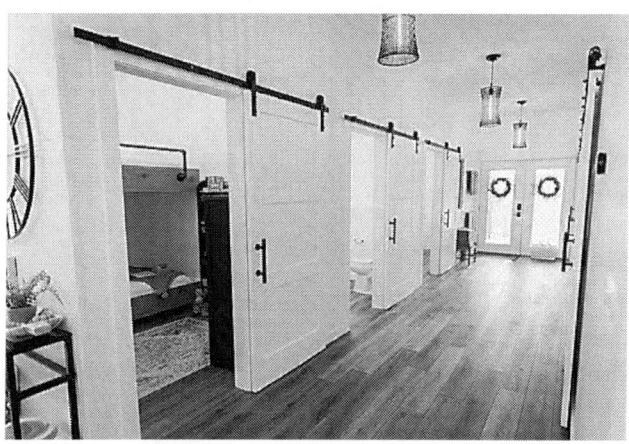

Access to all rooms from both inside and outside the rooms should be easy and unimpeded as much as possible for those aging in place for as long as possible. It is quite a necessity to have wide doorways and hallways when it involves making a home accessible to people using walkers or wheelchairs. As far as designing a floor plan for a barndominium, hallways should be at least 36 inches wide, while the minimum recommended width for doorways should be at least 32 inches wide to ensure access for all.

Of course, there are space-saving options in the way of pocket doors or sliding barn doors that greatly reduce the need for swinging doors, which can sometimes be tough to work around. Barndominiums are usually associated with a rustic style, in which these sorts of doors add not only style but more functionality to them.

- **The Comfy and Accessible Design of Bedroom Architecture**

The master bedroom should be well-planned with ease of access and comfort in mind and sit in a convenient place on the main floor. Allow an open area around the bed for the potential need for a mobility aid, caretaker, or future medical equipment, such as a hospital bed or lift chair. Bedrooms are to have abundant natural light because such condition aids in enhancing mood and visibility within the area.

Closets that are more accessible, for storage, have lower rods and shelves that make retrieval of objects such as clothes and personal effects easier to access. Walk-in closets may also be designed to be more accessible by wheelchair users through movable shelves and space for maneuverability.

- **Increasing Convenience and Safety Through Smart Home Technology**

This is where modern smart home technologies can play a very vital role in enhancing the safety and convenience of an aging-in-place barndominium. All these voice assistants, smart lighting systems, and automatic thermostats will go a long way in making day-to-day activities easier to conduct for a senior citizen.

For instance, one can install motion-activated lights in hallways and bathrooms to eliminate the risk of tripping and falls at night. Being able to see who's at the door without getting up creates great peace of mind for many homeowners, and smart video doorbells and security systems provide this opportunity. Moreover, the design of the home can include medical alert systems in such a way that they provide a direct link to emergency services, should these be necessary.

- **Accessibility of Outdoor Living Spaces Beyond the interiors**

Barndominiums may have extensive outdoor spaces that can be just as accessible inside as out. Wide and smooth walking paths, ramps where there is a set of steps, and raised bed gardens are ways elderly individuals can enjoy continued activities outdoors with no physical strain.

The outdoor patio should be covered, illuminated, and treated with a non-slippery surface to avoid any accidents. This helps the elderly enjoy the outdoors and the natural setting in comfort for their seating and shade and with ease of access to the interior.

ADA Compliance: Features include Wide Hallways, Low Countertops, and No-step Entrances.

Given that society is aiming for universal design and inclusion, accessibility of persons with impairments to living environments has become of paramount concern. The ADA sets significant standards in ensuring that public spaces and buildings are accessible in a manner that would make entry and navigation possible regardless of a person's differing physical capabilities.

ADA compliance can easily be added to floor plans in residential construction, but it has become particularly popular with barndominiums-a combination or marriage of barn aesthetics with the ease and comfort of a house to ensure the floor plans are accessible, utilitarian, and comfortable for all.

For many reasons, barndominiums are adaptive, inexpensive, and good-looking-end. These homes were originally designed to be barns or workshop buildings that can later be converted into living areas and can be customized in many ways.

Including ADA-compliant architectural features in the floor design of a barndominium is of great importance to landowners or occupants who have disabilities in their mobility or other bodily limitations. Large hallways, low counters, and no stairs to access the entrances are just a few features that greatly contribute to the fact these homes are not only accessible but functional for all individuals regardless of abilities.

What does it mean for a residential home to comply with the Americans with Disabilities Act?

The Americans with Disabilities Act is legislation enacted in 1990 that assures equal opportunities and access for individuals with disabilities. This is civil rights legislation enacted by the United States. While the ADA generally pertains to public areas or those used commercially, most of the concepts can be applied to residential design. This may be especially useful for one's immediate family or homeowners who have physical limitations.

These ADA-compliant homes seek to minimize physical barriers in an attempt to provide settings that will be accessible to persons with mobility impairments. Most of the design features in ADA master bedrooms help not only individuals with impairments but also elderly adults, parents with small children, and even able-bodied persons desiring added convenience and comfort in their residential settings.

The importance of wide hallways in navigable ways

Wide hallways are one of the most critical things a house should have to become ADA-compliant. Barndominium floor design with broad wide halls will contribute to significant changes as far as accessibility is concerned. The minimum corridor width of 36 inches is recommended by the ADA, and this would help people using wheelchairs, walkers, and other mobility aids get through easily. Wider hallways make navigating with furniture movers or families with strollers a great deal easier but it also allows room for accommodating the use of mobility equipment.

In barndominium designs, the rustic and open concept of the home lends itself to longer halls without necessarily sacrificing the aesthetic appeal of the design. Open-concept living rooms also tend to be integrated into the floor plan for a barndominium design. This has the potential to further increase accessibility by reducing the number of narrow doorways and tight corners. Wide hallways create a practical solution to mobility concerns when combined with non-slip, smooth flooring and mild transitions from room to room.

Low countertops with an accessible kitchen setup

The kitchen is said to be the "heart" of every home; therefore, it is also essential for every ADA-compliant barndominium to ensure that the kitchen is functional and accessible. Among the most essential characteristics that ought to be in an ADA-friendly kitchen is that of having low counters.

According to the Americans with Disabilities Act, it is recommended that worktops should not be higher than 34 inches high and there is knee space underneath them so that people in wheelchairs can access the work safely.

Other examples of ADA-compliant kitchen facilities that can be incorporated into barndominium floor plans, in addition to simply having below-average-height kitchen counters, include the following:

- **Pull-out shelves and drawers**: These features make it easier for people to access products stored in cabinets without having to bend down or stretch to the top of the cabinet.

- **More accessible appliances:** Mounting oven, microwave, and dishwasher at a lower height provides access for people with mobility disabilities easily. Ovens opening from the side and stovetop installed with front controls are ideally accessible.
- **Wide turning radius:** Kitchens should have a turning radius wide enough-at least sixty inches so that a wheelchair or other mobility device can make a complete turn.
- **Lever-handled faucets and controls that are easy to manage**: These features minimize the amount of grasping, twisting, or pulling that must be done; they are more convenient for someone with limited hand dexterity.

These easily accessible kitchen features make it possible for the barndominium to maintain the functionality and accessibility of the kitchen to all residents with any physical capability.

Eliminate entry barriers with no-step entries

Entry into homes comes with various complications for persons with mobility issues, but maybe the most daunting of those is having to enter over steps or unevenness. The removal or minimizing of the number of steps at all entry points will ensure that barndominium floor designs comply with the ADA.

Besides being a zero-threshold or zero-step entry, the entrance should not contain any steps. It should provide people with the capability to gain access to and leave the house without necessarily needing stairs or thresholds which may contain some complications in movement.

This characteristic in barndominiums can be realized with several design solutions including:

- **Sloped ramps and walkways**: You may want to avoid using steps up to the front door and instead use ramps or walkways with a gentle slope. The most recommended ramps will have a slope ratio of 1:12, meaning the slope must not be above one inch for every twelve inches of horizontal distance. The presence of handrails on both sides of the ramp further improves accessibility and safety.

- **Flush thresholds:** The doorways should have a flush threshold because it eliminates all possibilities of tripping plus one can move around easily without any difficulty most importantly for those using wheelchairs or walkers.

- **The wide entrances:** It simply means that besides not having steps, the doorways must be at least 36 inches wide to accommodate access for equipment in mobility and ease of passage for all occupants within.

No-step entrances not only cater to persons with disabilities but also make homes more accessible to elderly homeowners who may have difficulty in later years getting around. They make it easier for anyone to come and go from the house, a great convenience for parents who need to push a stroller or anyone who carries groceries and other cumbersome items.

Bathroom accessibility: ensuring individual safety and comfort

Some other important places the ADA dictates, especially for people unable to move themselves, are bathrooms. Following are some accessible features per ADA to be incorporated into bathroom design in barndominium floor designs for accessibility.

- **Roll-in showers:** A shower without a raised threshold enables users in wheelchairs and walkers to easily roll into the shower without any assistance. Other features that should be included with a roll-in shower to increase safety and convenience are grab bars, nonslip floors, and a handheld showerhead.

- **Comfort-height toilets:** These are toilets installed higher than the conventional height, normally in the range of 17 to 19 inches. Such a raised toilet is more accessible to persons with limited mobility.

- **Grab bars:** From the viewpoint of prevention and stability for those who have limited mobility, the installation of grab bars around the toilet and inside the shower/bathtub can be used.
- **Accessible sinks:** They provide an opportunity for wheelchair users to wash their hands in a comfortable position. Sinks with knee clearance underneath them. Also, faucets with lever handles or touchless options make the procedure of using them more convenient.

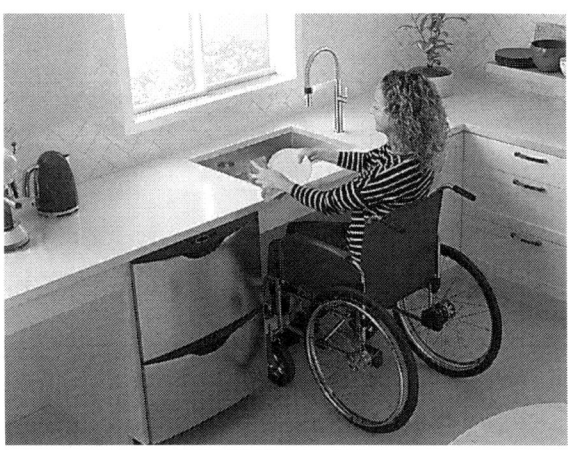

With these ADA-compliant features in its design, a barndominium can offer people with disabilities a safer and more pleasant environment.

Several Advantages of ADA-Compliant Barndominiums

Besides meeting the basic needs of persons with disabilities, ADA-approved barndominiums offer numerous advantages to homeowners. Designing a future-proof home will allow older homeowners to live in their homes longer and won't necessarily include costly repairs later on. Seamless open design in barndominiums lends itself well to accessible living, and it's not hard to put components of the ADA into them without sacrificing aesthetics or function.

Besides, ADA-compliant homes could be much more marketable, serving a wider range of potential buyers from families with elderly parents down to those who have disabilities. By incorporating accessibility into their initial design, homeowners will be able to create an environment that is not only accommodating but also functional and beautiful for many years to come.

CHAPTER TEN

Sample Barndominium Floor Plans

Visual examples and blueprints of popular designs

1. **Single-Story Barndominium Open Concept Living**

Total Square Feet: 1,800 sq. ft

Rooms: 3 bedrooms, 2 bathrooms

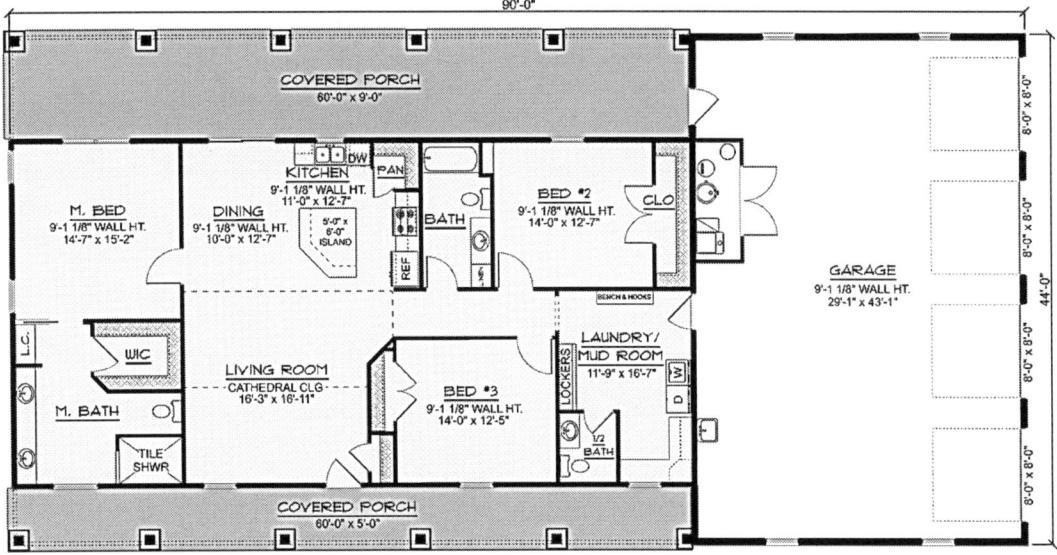

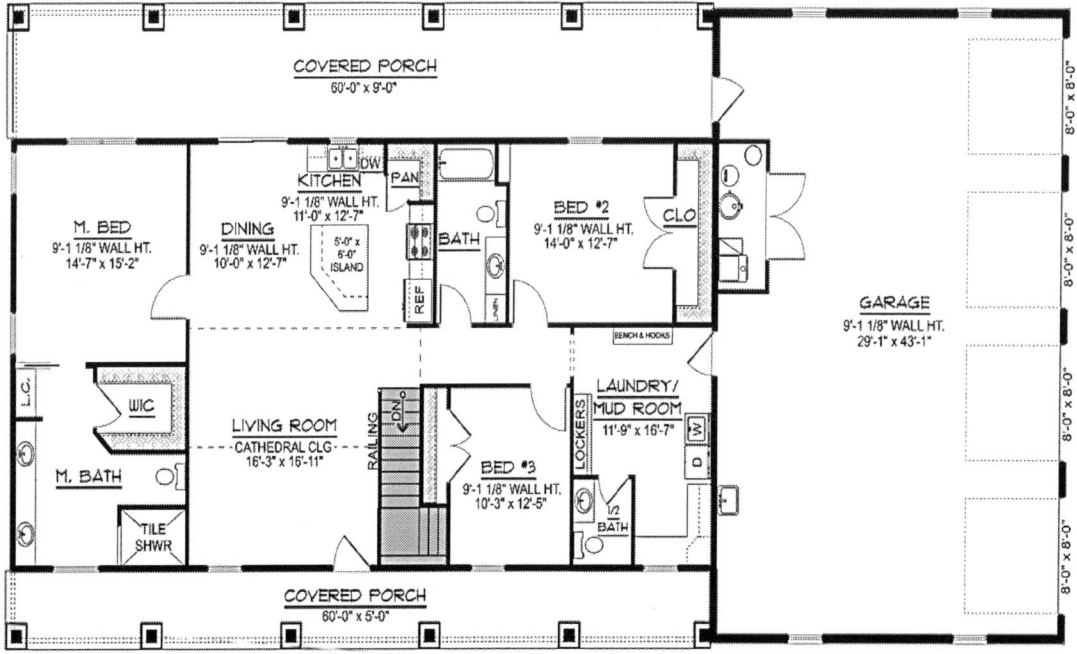

Layout Features:

- Open living room: Large open-concept living area open to the kitchen and dining space.
- Kitchen: Central island with seating, pantry, and ample counter space.
- Master Suite: Private master bedroom with a large walk-in closet, along with an attached en-suite bath with a double vanity and shower.
- Additional Bedrooms: Two bedrooms that share a full bath.
- Utility Area: Laundry room off the kitchen.
- Garage/Shop: Att'd 2 car garage w/workshop area.

2. Two-story Barndominium with Loft and Garage

Total Square Feet: 2,500 sq. ft

Rooms: 4 bedrooms, 3 bathrooms

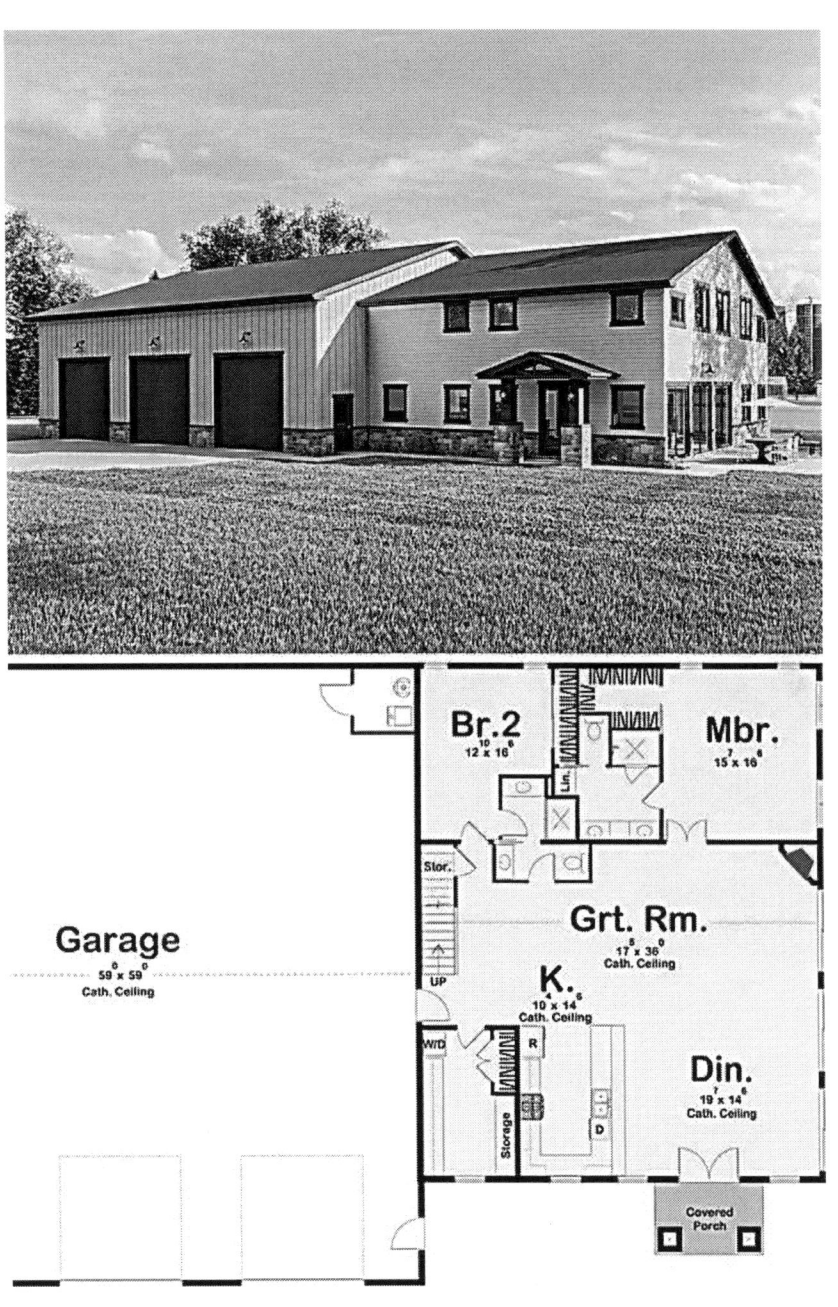

Layout Features:

- Living Room: Vaulted ceiling living room w/loft that overlooks
- Kitchen and Dining Area: Large kitchen w/island, breakfast bar, and dining area opening to the living room.
- Master Suite: First-floor master suite with a walk-in closet and an en-suite bathroom with a soaking tub and separate shower.
- Loft Area: Second-floor loft area perfect for a family room or home office.
- Other Bedrooms: Two upstairs bedrooms that share a full bath and one downstairs guest room.
- Garage: Three-car attached garage leads directly into a mudroom.

3. L-Shaped Barndominium with Wraparound Porch

Total Square Feet: 2,200 sq. ft

Rooms: 3 bedrooms, 2.5 bathrooms (two full bathrooms and one 'half' bathroom)

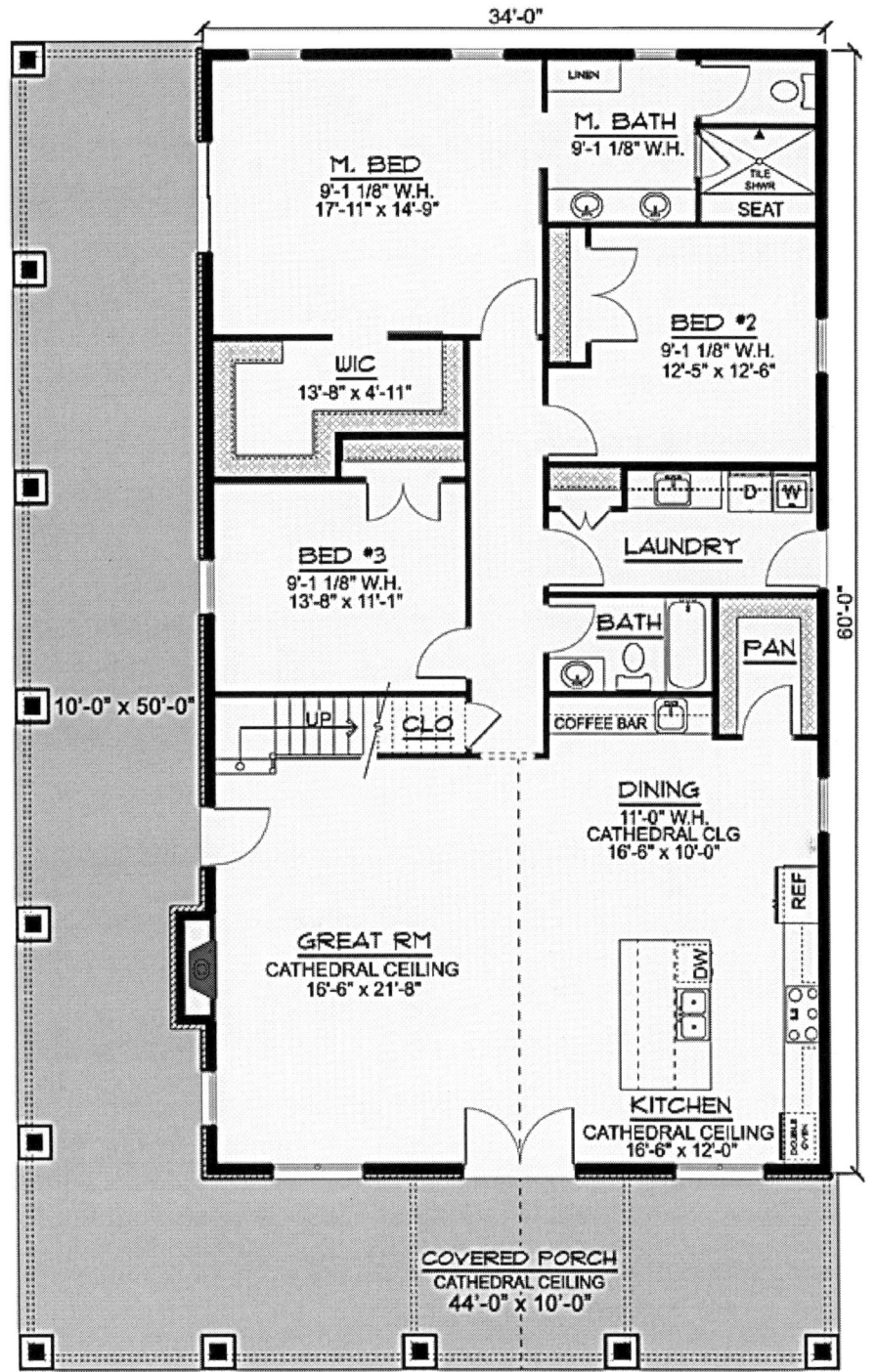

34'-0"

M. BED
9'-1 1/8" W.H.
17'-11" x 14'-9"

LINEN

M. BATH
9'-1 1/8" W.H.

TILE
SHWR

SEAT

WIC
13'-8" x 4'-11"

BED #2
9'-1 1/8" W.H.
12'-5" x 12'-6"

D   W

LAUNDRY

BED #3
9'-1 1/8" W.H.
13'-8" x 11'-1"

BATH

PAN

10'-0" x 50'-0"

UP

CLO

COFFEE BAR

60'-0"

DINING
11'-0" W.H.
CATHEDRAL CLG
16'-6" x 10'-0"

GREAT RM
CATHEDRAL CEILING
16'-6" x 21'-8"

REF

DW

KITCHEN
CATHEDRAL CEILING
16'-6" x 12'-0"

DOUBLE OVEN

COVERED PORCH
CATHEDRAL CEILING
44'-0" x 10'-0"

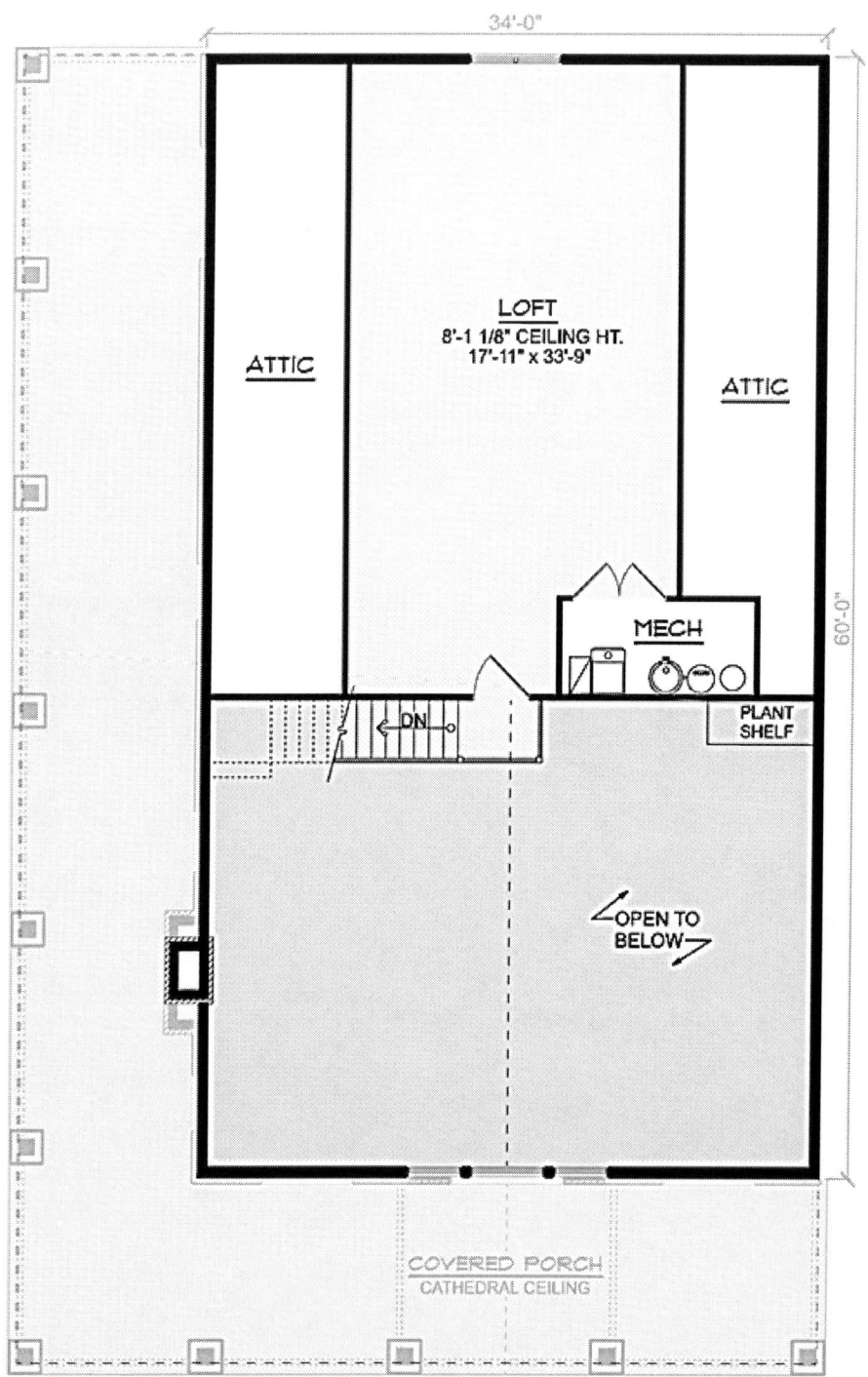

ATTIC

LOFT
8'-1 1/8" CEILING HT.
17'-11" x 33'-9"

ATTIC

MECH

DN

PLANT
SHELF

OPEN TO
BELOW

34'-0"

60'-0"

COVERED PORCH
CATHEDRAL CEILING

Layout Features:

- Wraparound Porch: This is a large, L-shaped porch, wrapping living and dining spaces for outdoor living.
- Great Room: Open great room with vaulted ceilings, ideal for family gatherings.
- Kitchen: Spacious kitchen with a breakfast bar, walk-in pantry, and plenty of cabinet space.
- Master Suite: A private master wing with a walk-in closet, en-suite bathroom with dual sinks, soaking tub, and shower.
- Additional Bedrooms: Two bedrooms share a bath.
- Laundry Room: This is located near the entrance of the garage.
- Bonus Room: Flex room for a home office or playroom.

4. **Barndominium with a Large Shop Area**

Total Square Feet: 3,000 sq. ft

Rooms: 3 bedrooms, 2.5 bathrooms, 1 large workshop

10'0" X 8'0" 12'0" X 10'0" 12'0" X 10'0" 10'0" X 8'0"

Slope

Shop
12' Sloping CLG.
50'4" X 24'9"

Slope

Laundry
5'6" X 9'0"

Bath
4'8" X 4'7"

Kitchen
10' Sloping CLG.
26'4" X 17'10"

10'2" X 10'0"

Master Bath
10' Sloping CLG.
10' X 11'4"

Closet

Covered Porch
10' Sloping CLG.
13'6" X 16'0"

Master Bed
10' Sloping CLG.
10'2" X 16'10"

Great Room
19'6" Sloping CLG.
29'10" X 17'8"
Open
to Above

Slope

Slope

10'2" X 10'0"

Covered Porch
10' Flat CLG.
26'0" X 8'0"

155

10'0" X 8'0" | **12'0" X 10'0"** | **12'0" X 10'0"** | **10'0" X 8'0"**

Slope →

Shop
12' Sloping CLG.
50'4" X 24'9"

← Slope

Laundry
5'6" X 9'0"

Bath
4'6" X

Bath 2
7'0" X 12'0"

Bedroom 2
8' Flat CLG.
10'8" X 14'0"

Bedroom 3
8' Flat CLG.
10'8" X 14'0"

10'2" X 10'0"

Master Bath
10' Sloping CLG
10' X 11'4"

4'-0"
CLO

4'-0"
CLO

Loft

DN

Covered Porch
10' Sloping CLG.
13'6" X 16'0"

Master Bed
10' Sloping CL
10'2" X 16'10"

Open to Below

Slope
SLOPE

Slope
←

10'2" X 10'0"

Covered Porch
10' Flat CLG.
26'0" X 8'0"

Features of the Layout:

- Large Shop Area: The attached shop area consists of 1,200 sq ft for working on projects or storing equipment.
- Living Area: Spacious open-plan living, dining, and kitchen area.

- Master Suite: This oversized master suite features a walk-in closet and a full bathroom.
- Additional Bedrooms: Two guest bedrooms, located on the other side of the living area, share a bathroom.
- Mudroom and Laundry: A connecting mudroom between the shop and house with laundry.
- Office/Flex Room: Additional space that can serve as an office or hobby room.

Case studies in different layouts

Case Study 1: Single-Story Open Concept Barndominium

Key Features:

- The Open concept living area: The living room, kitchen, and dining room have all combined in this area of the house into one large open area. This is the heart of the house. The large windows bring in a lot of light and also add to the openness and airiness of the room.
- Master Bedroom: The master suite sits on one side of the additional bedrooms with a walk-in closet and an en-suite to enhance the level of seclusion.
- Outdoor living: The rear of the house is wrapped with a porch extending across the full width, thus making this an exterior dining and lounge space accessible from the inside.

This one-story design was apt for a busy family looking for a modern and open space for recreational purposes and day-to-day family living. Maintenance chores were also easier to perform, and the building proved to be more energy efficient due to the design of this one-story edifice.

Case Study 2: A Two-Story Barndominium with Loft and Workshop

Key Features:

- Loft: Overlooking the living room, this loft may also act as a home office or amusement area because it is quiet yet connected at the same time.
- Workshop: A thousand-square-foot workshop located within walking distance from the main house. Due to its location, it is large enough to handle the owner's side woodworking business quite nicely.
- Entertainment Space: Open-concept floor plan that was developed in the kitchen, dining area, and living room to accommodate entertaining. The dining area off the island kitchen opens to the patio; high-end appliances grace the kitchen.

While the bedrooms were on the second level, the loft living area was completely private and offered an open living space for any type of use. The attached workshop allows cohesion to be integrated right from the office into the home, adding to the efficiency of the workshop.

Case Study 3: L-shaped Barndominium with Wraparound Porch

Key Features:

- L-Shaped Layout: This L-shaped layout lets the home embrace its surroundings, giving stunning views of the property from almost every room in the house.
- Wraparound porch: Featuring a wraparound porch, the porch is easily accessible from both the main living space and the master bedroom, making it an ideal location to sit outside and host gatherings.
- The Master Suite: It is to be found in one of the wings, and has a very huge bedroom, a walk-in closet, and an opulent en-suite bathroom with a soaking tub.

The design called for an L-shaped footprint for the property that would allow the owners to leverage the rural setting of the property and create multiple options for outdoor living. There is nothing better than being able to enjoy the breathtaking scenery from the large porch that is already rapidly becoming their favorite feature in the house.

Case Study 4: A Barndominium with a Large Shop for Hobbies and Other Activities

Key Features:

- Workshop: Owners can enjoy their hobbies of restoring classic cars and woodworking with the more than 1,200-square-foot workshop adjacent to it. It has a separate entrance and bathroom, hence considered self-contained.
- Primary Room: The main or primary room is centrally located on the ground floor of the house. It is comprised of an open floor plan, which covers the living room, dining area, and kitchen area. A large patio and garden come into view through the opening of the room
- Segregated Bedroom Areas: The master suite will be on one side of the house, while the guest bedrooms and children's bedrooms will be on the other side. This would mean that no matter how many people come into the house, the residence remains quiet and private for the owners.

Results showed that the mix of a living space and retail area was about perfect for a couple's way of life. The open-concept living space proved quite an asset regarding family get-togethers, and the spacious workshop quickly became an "ultra favorite" feature.

Each of these case studies shows how a barndominium plan might be constructed for a homeowner in various ways, including with open living spaces, connectedness to outdoor living spaces, or a workshop area that serves a practical purpose.

Pros and cons of various designs

After considering different types of Barndominium floor plans, you will have to review the pros and cons involved in every design. Your needs, likings, and way of life determine this question. Below is an outline of all the different designs, together with the advantages and disadvantages of each.

1. Single-Story Open Concept Barndominium

Pros:

- Ease of Accessibility: Because all the rooms are on the same level, it is very convenient for those who have difficulties in movement, such as children, elderly people, or any other person with mobility problems.
- Open Layout: This type of layout is designed to be quite spacious, and because of that, it allows family members to interact and enjoy themselves.
- Energy Efficiency: A single-story house is generally more efficient in terms of cooling and heating since the air can be circulated consistently through the house. This enables heating and cooling more efficiently.
- Low Maintenance: It's just that it's much easier to maintain because there are no stairs or other floors. This is one of the advantages of this design.

Cons:

- Less Privacy: Since it is an open architecture connected, it generally allows for not much privacy.
- Smaller Scope of Expansion: If your family grows in size or you need more rooms someday, you will be expanding laterally, and the scope of the expansion will be much smaller. When you want to expand your single-story home, it involves a larger section of your lot.
- Noise not Consolidated: Since this is an open building, the noise from the kitchen, living room, or other shared areas may easily find its way into your home.

2. Two-story Barndominium, Including Loft and Workshop

Pros*:*

- Living Space: The two-story houses give you more square feet of living space without the extra cost of a larger square footage property. Because of this, it is an excellent alternative for huge families or to extend further rooms for guests or hobby purposes.
- Private Loft Area: Some characteristics of the loft include a private area that can be utilized for a family room, game room, or office.
- Attached Workshop: Because it is attached to the house, it could serve practical hobby purposes, businesses, or extra storage places. Separation of Living places: The bedrooms or other quiet places would be upstairs while the social ones are downstairs.

Cons:

- Not as accessible: Stairs can be a pain, especially for people who are elderly or who do not feel as though they have sufficient mobility.
- Higher heating and cooling costs: Because heat rises, temperatures in two-story homes may not be evenly distributed, and this will make the amount of money spent on energy to maintain comfort higher.
- Exterior Second Story Maintenance: Normally, window cleaning or any kind of repair on the second story's exterior is not as easy and is costlier than the same work on the first level.

3. A barndominium Featuring a Wraparound Porch and L-shaped Floor Plan

Pros:

- Outdoor Living: This porch wraps around the entire house, providing much space for outdoor living to dine, entertain, or simply relax.
- A Private Master Wing: Sometimes, when the L-shaped nature of the floor plan allows it, the master suite can be set inside a private living space, making it even more secluded.
- Easy Incorporation of Indoor and Outdoor Living Space: Incorporating indoor and outdoor living space can be made much easier when there is a porch, along with an open floor plan. It helps in giving the two spaces a seamless transition from each other.

Cons:

- Complex Rooflines: The rooflines of L-shaped homes are usually more complex compared to other types of houses, increasing their construction and maintenance costs.

- A Waste of Potential Space: It would depend on the structure, but this L-shaped structure could lead to some areas of the house not being used as much, or it might bring discomfort in putting furniture.
- More Lot Size: Typically, one has to have a larger lot to appropriately appreciate the full value of an L-shaped layout and all that an outdoors has to offer.

4. A Barndominium with a Considerable Garage Area

Pros:

- Attached Shop: The attached shop is large, providing a separate workspace for working, hobbies, or storage without using any of the living space in the house.
- Separate Entrance: Most workshops have a separate entrance through which your clients or visitors may enter your work area without necessarily having to pass through the house.
- Multifunctional: This best describes the fact that it can later be used for other purposes like being converted to a garage or creating room for expanding living space.

Cons:

- Noise and Odor: If the store happens to be right next to where one resides, there are chances that noise from the machinery or the odors from the work may seep into the quarters.
- Cost: The cost of the workshop would add a considerable amount to the overall budget for the house, depending on how big the size of the equipment added would be and the quality of the equipment.
- Wasted Space: If it is not put to proper use, it has the potential to become just another one of those storage spaces that serves no actual use in terms of practical life.

5. U-shaped Barndominium with a Central Courtyard

Pros:

- Private Courtyard: The center courtyard converts the rear grounds into a somewhat private space that is screened from the wind and provides a protected spot, thus decreasing the neighbor's line of sight into the premises.
- Perfect for Parties: There are plenty of birthday parties the courtyard model would be ideal for providing an open area outside the house, with several rooms opening onto it.
- Natural Light: Large windows facing on all sides will bring natural light inside the home and enhance the energy efficiency of the courtyard.
- Versatility in Design: This allows one to easily segregate the public from private areas and makes it highly versatile in accommodating all sorts of different family configurations.

Cons:

- More Exterior Walls: This will lead to one having exterior walls, which will then mean building costs will be higher, as will the maintenance requirements.
- A Lot-wasting Design: The courtyard will consume the major area of the lot depending on the lot form, leaving less area for other outside activities.
- It is a More Complicated Construction: Based on the difficulty this structure has, it can still be more expensive and will take a lot of time to build.

All these decisions depend on various factors like your lifestyle, size of the lot, budget, and for what purpose it is being used. Every aspect of these designs has its positive and negative sides.

CHAPTER ELEVEN

Tips for Choosing or Creating a Barndominium Floor Plan

Working with Professionals: Architects and Designers

Collaborating with architects and designers is necessary to make sure your dream comes into reality. In this section, we go over how to engage the pros, specifically architects and designers, to bring the floor plans of your barndominium into reality and how you can best use your collaboration.

1. Understanding the Role of the Architects and Designers

The detailing of engaging the professionals requires understanding the difference in roles between architects and designers.

- A licensed professional (an architect) who has the training to design buildings, keeping in view both the aesthetics and structural stability of the building. Apart from ensuring that your barndominium is structurally solid, they make sure it complies with local building rules and safety norms.
- Design professionals, on the other hand, are more concerned with the aesthetics and layout of the inside. These professionals' services are primarily required to make the interior space more beautiful and user-friendly; nevertheless, many of them find themselves being pulled into the architectural element of the building, as well. Though not necessarily licensed, designers know it all, especially on how to optimize space utilization, what materials to use, and how to have your barndominium be your expression.

They play a vital role in planning and executing the carrying out of barndominium floor plans, respectively. The cumulative knowledge these individuals possess assures you that your room will be attractive, functional, and safe.

2. Setting the Stage for the Future: The First Consultation

The initial consultation would be the process of working with an architect or designer. This phase allows one to share his vision regarding the development of the barndominium. One should go ahead and communicate his needs, lifestyle, and requirements particular to himself during this phase. I would want to address the following main points:

- **Purpose of the Barndominium**: The Barndominium is a space of many functions such as the house, office, shop, or storage room for equipment or animals. Are you building a family home, a retreat from the world, or a place with combined living and the above?
- **Size and Layout Preferences**: How many bedrooms and bathrooms does one need? Is an open-concept design living area preferred, or are there those who like places compartmentalized?
- **Specific Features**: Would you like a loft, huge windows to let in lots of light, wraparound porches, or energy-efficient elements, such as solar panel installation or materials providing a very high level of insulation?
- **Budget:** Developing a budget from the very beginning will help both you and the professionals stay in perspective on what is feasible. In addition, it makes certain that the architect will not design something that is far too expensive for you to be able to pay for.

Your architect or designer will take this information and use it as a starting point from which to develop a design that matches your ideal picture. Since this consultation is going to set the stage for everything else, it should be as candid and specific as possible in this preliminary phase of the process.

3. Schematic Floor Plan Development

Once the meeting is complete, the architect and designer will start working on the floor plan of your home. This would include creativity and functionality since it has to meet your space needs and style.

- **Maximize Space:** The floor plan will try to maximize space availability. Since most barndominiums have open floor plans, your architect has to work out how to arrange the spaces with much flow: living, kitchens, bedrooms, and other areas. Interior layouts on interior spaces- designers ensure furniture arrangement, walkways, and storage options make sense.
- **Bringing Form and Function Together**: While most barndominiums are relatively simple in design, it takes a little creativity to meld the charm of a barn with the comforts of today's lifestyles. When it comes to tying the design together, your architect will consider exterior elements like rooflines, windows, and porches, while your designer may have suggestions regarding finishes, color, and texture.
- **Sustainability and Energy Efficiency.** Many owners building barndominiums today have an interest in green building principles, especially those having to do with sustainability and energy efficiency. Appropriately insulating the structure, placing windows to optimize passive solar gain, and using environmentally friendly building materials are some ways your architect can add energy-efficient design principles. Eco-friendly coatings and smart appliances are two options that designers could suggest.

You will most often be discussing the preliminary floor plan that has been developed with both your professional architect and designer. Through this review procedure, you will still be allowed to make modifications or comments before the final design is completed.

4. Considering the Zoning Laws and Building Codes

The most important duty that the architect has is to ensure the floor plan for the barndominium is compatible with all local zoning laws and building codes. Every county and city has its set of rules and regulations to which homebuilders must conform while constructing. These become even more complex with the construction of non-traditional dwellings like barndominiums.

Your architect would be responsible for ensuring the structure meets the following needs:

- **Zoning Restrictions:** You might have to deal with specific land use limitations in the county or municipality you are developing in. Several areas put limits on the types of structures that can be built on agricultural or rural property although other areas have building size limitations or footprints from the ground up.
- **Building Codes:** These guarantee that your project is safe and enduring during construction. This would come under the responsibility of your architect to ensure that the structure, electrical, plumbing, and other essential systems meet the standards of these codes.
- **Special Permits and Approvals:** Most areas or regions have special permits that need to be issued before any new structure can be put up. The application process will be mainly handled by your architect, further smoothing the process of your barndominium project.

It could entail some highly expensive delays or, worse, have parts of your work redone, depending on the severity of your mistake in case you fail to pay attention or simply ignore zoning rules and construction codes. Because of this very fact, hiring an experienced professional architect is mandatory.

5. Collaboration during Aesthetic Decision-Making

After an agreement is reached on the structural design and floor plan of your barndominium, you will closely work with your designer in the selection of the aesthetics of the finished product. In that regard, you will be responsible for coordinating with your designer to make selections regarding interior layouts, colors, materials, and finishes compatible with your unique sense of style. The following represents some areas where designers add value:

- **Material Selection:** The selection of materials is one service that designers will offer to assist you in making the right choice of materials for wall treatments, cabinetry, worktops, and flooring. Durability, ease of maintenance, and how well the materials fit into your overall vision are factors they consider.

- **Color Palettes:** Your designer will be able to help you choose color palettes that enhance the architecture and add ambiance to the spaces you may have. They can assist you in picking colors, furniture, and other elements of décor that best suit your tastes, whether modern, rustic, or eclectic.
- **Furniture Arrangement:** Besides helping you choose furniture, designers also optimize furniture arrangements to create comfortable and functional areas. In a barndominium that usually contains open floor plans, furniture arrangement is one important factor in determining many zones contained inside the area.
- **Lighting & Fixtures:** Lighting is a major design element that can change the feel of a place. The lighting options that your designer will propose are those that will maximize the space and emphasize the main architectural features.

6. **The Importance of Communication in the Modern World**

Communication with the architects and designers is continuous from the very start to the end. Setting up open lines of communication ensures your expectations are met and any potential problems are recognized early on in the design process. You will need to make provisions with your pros for frequent check-ins and review sessions.

- **Understand your Vision**: Make sure you understand what your vision is, and do not be afraid to ask questions or state your concerns if something does not align with your goals.
- **Give Prompt Feedback**: Give feedback as soon as possible because the sooner you catch something that needs to be different, the easier it will be to make those changes.
- **Trust their Experience**: Years have been spent by an architect and a designer in the industry, and most of the proposals are usually supported by industrial wisdom or practical issues. As much as you may have a very good vision, allow yourself to remember that they have experience.

Off-the-shelf Plans Versus Custom Plans: The Positives and Negatives of Each

The first major decisions to be made at the beginning of the building process of a barndominium are between pre-designed floor plans and those designed with the particular project in mind. Each of these two options has its potential advantages and disadvantages, especially depending on taste, need, or budget. Each of these avenues is laced with its share of pros and cons, which this essay will tackle in due course to assist you in making up your mind about your barndominium business.

Barndominium Off-the-shelf Floor Plans

Off-the-shelf floor plans, also called pre-designed or stock floor plans, are designs that have already been engineered and are for sale. These barndominium building plans are normally prepared by architects and designers who have a professional license in the field. They are also designed to accommodate several tastes and often offer a compilation of different floor plans, square footage, and room arrangements. Let's have a look at both the pros and cons of using pre-made plans instead of making them yourself.

Pros

- **Affordability:** One of the great benefits when it comes to using pre-made off-the-shelf barndominium floor plans. Off-the-shelf designs have several advantages, the most significant of which deals with cost. Since these designs have already been created, they are considerably cheaper than if an architect or designer were to be hired to create one from scratch for a custom project. You can save thousands of dollars, depending on the level of complexity and the size of the floor plan. For many first-time builders and those on a tight budget, this is quite an incentive. Additionally, many pre-designed plans have structural and technical specifications that meet pre-approval from building departments. This further reduces the cost of changes or compliance,

- **Efficient use of time:** The process of building a barndominium is mostly appealing but time-consuming. Since the designs are readily available, you rush the planning phase once you adapt over-the-counter blueprints. There isn't any necessity to invest weeks or even months of going back and forth with an architect just to create a strategy that's specific to your project. Once you identify a plan that best suits your needs, construction can begin. That is a convenient and time-saving feature, which may be highly desirable for people who want to move into their barndominium sooner or simply cannot stand waiting to complete their project.

- **Full range of options**: Prefabricated floor designs are available in a large variety to suit almost every taste and preference. Whether it's a small, compact, and minimalistic design or a large, sprawling concept with multiple rooms, there may be a plan to suit your needs. Most businesses offer a full catalog of designs in which room dimensions can be deliberated alongside the number of bathrooms and added storage among other attributes. While some changes need to be made, such diversity is a great start for a homeowner.
- **Designs that have been tested:** The good thing about pre-designed plans is that they have more often than not undergone quite a few rounds of testing and building processes, which validates the fact that the blueprints work well and are practical. Builders can also comment on what they have to say about the building process for these plans, after which the constructions can be modified over successive periods. That implies that the off-the-shelf plans have already been tried in reality, which takes away most possibilities of design flaws or structural inefficiencies arising.

Cons

- **Relatively low level of personalization:** One of the disadvantages of pre-made barndominium floor plans is a relatively low level of personalization. Although some plans allow for minor changes, such as replacing one bedroom with a home office, large changes can be difficult or expensive. Since stock plans are designed to appeal to a large number of people, they may not be tailored to your individual needs or tastes or to the site itself. For instance, if you have an odd lot shape or you have some specific design elements in mind, you might feel that those pre-made plans are just too confining for your situation.
- **Lack of customization:** Some owners feel like one of the primary reasons to build a barndominium is to have a home that is truly unique home that shows who they are and how they live their lives. Off-the-shelf designs, by their very nature, cannot allow the same level of customization as custom designs can. This could make one feel that their house isn't any different from some other copycat version, assuming, of course, you have a clear vision of what your space should look like and feel like you have some very specific needs that don't usually fall under most pre-drafted plans, like massive workshops or unique entertainment areas.
- **Incompatibility issues**: Some of these stock plans may not meet the local building codes or your site specifics. These stock plans are pre-designed, meaning they never designed your site when these plans were created. Major modifications may be required because of conditions like sloping land, unusual soil, and local requirements. This can wipe out some of the cost savings that you achieved. Sometimes, an existing plan will be a good option; however, there are occasions when modifications are problematic, time-consuming, and costly.

Custom Barndominium Floor Plans

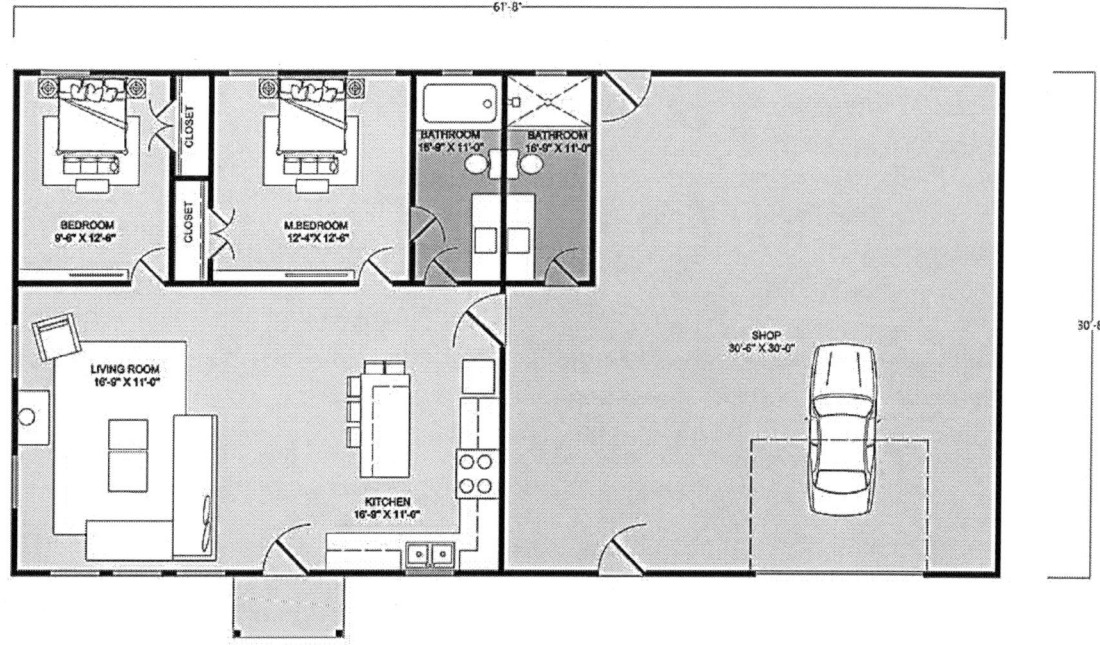

As the name suggests, custom barndominium floor plans are designed from scratch with your exact preference, needs, and specifics of your property in mind. In working with an architect or designer, one gets to create a home designed specifically according to preference. The following are the pros and cons of taking the custom route:

Pros

- **Customized designs**

Numerous advantages are associated with owning a custom barndominium floor plan. These include customized designs, among other benefits. The opportunity to design a home that is perfectly fit for your lifestyle is the major advantage of having custom blueprints drawn. With the barndominium, full control over each aspect of the structure is possible, from the number of rooms to the size of those rooms, to the orientation the building will have on your property.

Whether one requires a very large garage to serve as a workshop, very high ceilings, or some sort of custom living space with an open concept, customized plans can give all of these sorts of capabilities. This ability to personalize even goes to the minute details to allow you to input your preferences in product design.

- **Increases Potential of the Property**:

Custom plans give one the ability to maximize the unique features of your property. If your lot is on a hill, near a lake, or has other unique features, then a custom design will make sure the home takes full advantage of that. An architect can design the floor plan of your home to incorporate views, sunlight, and natural drainage, all of which enhance the beauty as well as the functionality of your home.

- **Planning for the Future**

With a custom design, you can plan for your future needs. A custom plan lets you incorporate things that might otherwise be missing if you were choosing from off-the-shelf plans: planning for an expanding family, aging in place, or even future resale value on your property. Energy-efficient inclusions, flexible living areas, and accessibility features that may be incorporated into your barndominium hold great promise for its adaptability in response to changes in your needs.

- **Freedom to be Creative**

While constructing a barndominium that is personalized to fit your needs, no constraint is imposed on you from the pre-designed plans. You are free to let your imagination soar, explore rooms arranged in non-conventional ways, introduce new design features, and use materials made to order. It's only confined by your imagination and budget, which also makes it possible to plan a truly unique house.

Cons

- **Higher Price:**

Whereas an off-the-shelf design may be relatively inexpensive, custom floor plans will be significantly more in price. Adding up the cost of hiring an architect or designer, multiple revisions, and special features can drive up the upfront cost much higher. Sometimes, a custom plan can run double or triple in cost compared to pre-designed plans. More specifically, construction may go up in cost since a unique design could require specialized materials or ways of construction.

- **Time-consuming:**

Making a floor plan is a time-consuming job, and it is going to take a lot of liaising between you and the architects, revising designs a million times, and perfecting every single minor detail. Depending on its complexity and the pace with which decisions will be made, this may very well stretch into some months by itself and affect the actual making of the house.

- **Risk of Design Flaws:**

Because custom designs are one-of-a-kind, there is no real-world feedback to guarantee that every part of the plan will work perfectly. While architects and builders can anticipate most potential problems, there is always some level of risk involved that certain elements of the design will not function as well in real life as expected. This may lead to very expensive fixes or adjustments when construction is well underway.

- **More Complicated Approval Processes:**

One of the reasons custom plans take more time is that they require more complicated processes for approval with local building authorities. It could be that off-the-shelf plans are pre-approved or easily adapted to meet codes, while the same detailed review in the case of a custom plan extends the time taken for approval, which in turn adds to the overall project timeline.

Common Mistakes Made: Pitfalls in Plan Selection or Plan Development

Homeowners who embark upon planning and developing a barndominium tend to make certain common mistakes when it comes to choosing the right floor plan or developing the floor plan. It eventually provides homeowners with an immensely practical living space, visually pleasing with open-concept and other unique features.

However, since these structures are a bit unconventional, some design flaws could drastically affect the outcome. Below are some of the common mistakes and pitfalls you should try to avoid when choosing or designing a floor plan for a barndominium.

A. **Not realizing the importance of planning and preparation**

Perhaps one of the biggest mistakes made in the process of selecting or developing barndominium floor plans is not taking part in the full and detailed planning. Many owners get so excited about the opportunity to build a custom home that they dive head-first into the process without considering all the different elements of construction. This huge planning if skipped, might lead to problems that arise unnoticed after some time, such as room sizes that are not functional or proper layouts, or even structural problems that entail costly additions and modifications.

While carrying out intensive planning, needs in the past and those in the future should be considered. Homeowners need to consider not only the immediate long-term living conditions but also the urgent wants. A family who is planning to grow up in the future may need additional rooms or an extended living space, while retirees may be interested in accessibility and ease of maintenance. Otherwise, they may be planning to remodel sooner rather than later in their lives if they don't have such foresight.

B. Inability to acknowledge budgetary limitations

Another common mistake people make is that they underestimate the costs that are associated with a custom barndominium. The affordability of a barndominium compared to regular homes is often the selling point for many house owners, but it is wise to note that other costs can emerge out of nowhere. Custom materials, enhancements to structural elements, or specifications due to the local building code are some of the kinds of costs. Without an accurate, realistic budget up front, the homeowner risks running out of money before the project is complete. This could leave rooms unfinished or finished using lower-quality materials to save money.

It is important in a barndominium floor plan design to take into consideration the eventual cost of the construction and the additional costs that would result from offering customization options. There needs to be a very clear vision regarding the pre-construction costs involved. This includes labor costs, materials, permits, clearing the site, utilities, and other unexpected charges. Having a contingency is quite critical in covering any unexpected costs that might come up during construction.

C. Poor consideration of flow in the floor plan

A floor plan's flow describes how rooms and spaces inside a house will connect, and with what ease persons may be able to move around the house. Arguably, one of the biggest positives in a barndominium comes from an open-concept design, but when laid out improperly, this can lead to inefficient layouts that hamper the flow of everyday living.

The typical mistakes include inconvenient room locations. For instance, owners who consider proximity a key factor may consider it less desirable to have the master bedroom away from the kitchen or living area. Situating the washing room on the other side of the house where the bedrooms are, will significantly increase the amount of extra discomfort caused in day-to-day activities.

The placement of utility rooms and features like the HVAC systems, water heaters, and storage is also very important. These places should be put in positions where noise pollution will be minimal and convenience will be maximized. An HVAC system that is positioned next to a wall in a bedroom could lead to noise disruptions.

D. Not considering natural light and ventilation

Not being able to realize the importance of putting natural light and ventilation into the barndominium floor plan is one of the other general mistakes that are made by most homeowners. A well-thought-out design for the floor plan will maximize natural light and airflow in a house, consequently reducing the need for artificial lighting and boosting energy efficiency.

Large windows and open spaces, or correct orientation towards the sun, can help greatly improve the aesthetics and comfort levels inside homes.

With less careful design, however, a barndominium could result in a dark and stuffy atmosphere. This would most definitely be true if too much space is allowed for the storage spaces or huge walls with no or too little window space. The installation of skylights, large windows, and adequate ventilation will help create an atmosphere that is bright, open, and airy.

E. Disregard to zoning law and building codes

In creating the floor plan, there is a tendency to ignore the various zoning rules and building codes. There comes a time when local regulations will hinder the project dimensions, design, and placing, and also the types of materials to be used. Even though barndominiums are more adaptable in comparison with conventional residential, they are also ruled by such standards. Overlooking these could end in costly fines, delays, or even totally revising the project.

Before selecting a barndominium floor design or drafting one, clients need to know the local laws regarding zoning, construction, and permit processing. One can avoid delays and also ensure that the project meets all the relevant standards by consulting with a local contractor or architect who is conversant with these restrictions.

F. The open space isn't exploited enough

The big open spaces of a barndominium allow many modifications, and these are generally cited as one of the most attractive features of this type of dwelling. Other homeowners, on the other hand, do not maximize such spaces and therefore come up with designs that are unimpressive or not optimized. For instance, open-concept living rooms tend to be a little too open, giving environments that feel stark and cold.

A balance must be achieved between open space and defined living areas. It's possible to define different areas within a barndominium by incorporating design elements like partial walls, furniture grouping, or even trick room dividers into the space without losing the open and airy feel that it allows. Also, storage options are something a homeowner should pay closer attention to, given that open floor plans normally force one to create creative ways of keeping clutter hidden yet still making the home attractive.

G. Aesthetics comes first, not functionality.

It is quite easy to get charmed by the aesthetic appeal a barndominium gives off, considering the multi-faceted world of types, finishes, and design features available. As much as beautiful aesthetics might mean, it should not come before usefulness at all. For instance, as appealing as the inset of huge windows from floor to ceiling may be from an appearance standpoint, unless they are properly insulated or installed about their directional orientation, they are only apt to result in undue heat loss or heat gain that will only increase energy costs.

Functionality will always have to come first about the layout, materials, and orientation of the barndominium. It is quite futile to please the eye above everyday usability, especially as far as the design of the house goes since a well-designed home is beautiful but practical, too.

H. Not considering the requirements for storage

Many people like expansive and open areas, and that's why they are into barndominiums, which doesn't mean that space for storage should not be considered. Not building adequate storage in the house is one of those mistakes taken for granted. Living rooms may be filled up with clutter, especially when there aren't any built-in storage facilities like closets, cupboards, or shelving across the whole floor plan.

Planning storage space is crucial to help keep the home in a working and clean state. Loft spaces utilized, storage built under stairs, and huge closets installed are all creative storage solutions that can help keep living areas clutter-free while maximizing available space.

I. Underestimation of a flexible future

Last but not least, one of the most common mistakes homeowners make involves their failure to consider future flexibility regarding floor layout. Life conditions may change, and the house that is perfect for a couple today may not be perfect for a growing family or for people who might want to stay in the same house.

Designed to include flexible elements such as extra bedrooms, multiple-use areas, and ease of adaptation for later changes, it allows the house to remain functional and comfortable over time. Besides saving money and avoiding headaches later on, there are other advantages when one is trying to plan possible future expansions or modifications at the initial design stage.

Changing Plans: How You can Take any Design and Make it Your Own

You don't have to take generic floor plans when it comes down to constructing the barndominium of your dreams. Probably one of the most attractive features of the barndominiums is that they offer infinite numbers of customization, which allows you to make the space fully personalized.

In this section, we discuss how you can modify any barndominium floor plan for your needs, preferences, and personal tastes. We will walk you through the key processes involved in building your dream home, be it an open-plan living area, more secluded quarters, or perhaps a mixture of rustic and modern elements. We do this from start to finish.

A. Gaining a Basic Understanding of the Barndominium Design

The proper way to understand the personalization of your barndominium starts with gaining a basic understanding of the design and structural composition. Typically, the floor plan of a barndominium includes spaciousness, often like those wide and open spaces evident in a barn.

All barndominiums are significantly a combination of living quarters in the form of bedrooms, kitchens, and bathrooms, added to workshops or garages. More so, high ceilings make up part of what constitutes barndominiums; they not only give an impression of a big environment but also make it possible to incorporate loft rooms, skylights, and other elements.

Since barndominiums are made from a lot of metallic or steel framing, they offer so much versatility in terms of design. Without load-carrying walls, flexibility concerning reworking layouts of the floor plan is maximized for adjustments that will yield exactly what you want.

B. Identifying Your Preferences and Needs in the Situation

This involves the determination of exact requirements and preferences. You need to think about how you are going to make use of the space-whether it will be a permanent dwelling, a holiday home, or even a combination whereby your home also functions as your place of business. Some of the questions that should be considered include:

- How many bedrooms and bathrooms you would require? Of course, it depends on your family size and how much space you will need when guests are coming over.
- What type of kitchen layout works best for you? Would you like a separate, private kitchen, or would you want an open kitchen with a huge island?
- Or, would you rather have more room for your job or some of your hobbies? It is possible to include an office, workshop, or studio in your barndominium design.
- What do you think is the ideal ratio of open areas to private areas one would wish? The idea appealed to some open, airy living areas while others like demarcated areas offering their share of privacy.

Once you have a clear vision as to your needs and wants, you will be able to look at any floor plan and know exactly how to adapt it to make it more compatible with your lifestyle.

C. Changes to the Floor Plan to Suit the Desired Layout

Now that you have decided on what you need, it is time to begin the process of modifying the floor plan for your barndominium. Since they are templates, most floor plans available online or through builders may be modified in several different ways.

- **Changing the room layouts**

This is one of the most common ways a floor plan can be customized, and it has to do with changing the layout of the rooms. You may find that some rooms are too small or too large for your needs, or perhaps the flow of the space simply does not align with how you live your life.

For instance, if you're an individual who enjoys entertaining, you may want to extend the kitchen and living area for more open space conducive to people and entertaining. On the other hand, in case you value privacy above everything, then you may rework the bedrooms and baths such that they fall in an entirely different location from the major living areas.

The location of certain rooms can be changed to take in the sun or the view. A master bedroom placed on the east side can awaken its occupants with the rising sun. Common rooms can be placed on the west side of the house to take advantage of the setting sun.

- **Adding or removing walls**

With the barndominium being an open frame set up in architecture, adding and removing walls can be easy. In case the floor plan you are working with feels like it may be a bit too constricted, then you can try removing a few walls so that the area could have more openness. You can also construct walls to accommodate places that you may want to separate, for example, your bedrooms, office, or playroom if you are the kind looking for privacy or specific areas.

If you happen to have, for example, an open kitchen and living space but would appreciate some partial separation of the two rooms, you could install a partial wall or a sliding barn door that would shut to create a personal atmosphere.

- **Incorporation of loft spaces into the design**

One of the distinctive features in the architecture of barndominiums is the high ceiling, which makes loft spaces within the structure feasible. Lofts work best in optimizing space vertically and may be used to serve various purposes: converting them into extra bedrooms, offices, or storage rooms. If the floor plan you're working with doesn't already include a loft, adding one in to create extra square footage without having to increase the footprint's size is very easily done.

They can also serve as multipurpose rooms; for example, a loft home office/guest room or a reading elbow that looks out into the living room comfort zone.

D. Making Exterior Changes

Even though most of the attention is usually geared towards the interior design of your barndominium, you would not want to lose sight of the fact that it outside makes this home no less important in making the home feel like it is yours. A typical barndominium outside design features an integration of features of rustic barn architecture with modern or industrial finishes; however, there is a lot of room to play around with the style.

- **Choosing appropriate materials and finishes**

The exterior materials that you select to put on your barndominium will make a huge impact on the appearance and texture of your building. You can go traditional barn feel using wood siding or metal panels. You can mix some modern elements into your building design and still have stone or brick accents. Whether you want a traditional red barn look, or a clean, neutral palette, you can make the color scheme your own choice in finishes is yours, and can be personalized to suit your taste.

- **Creating living spaces outside the home**

Consider adding outdoor living spaces to your barndominium design if you are an outdoorsy person. It could be anything from a covered porch to a wraparound deck or even an outdoor kitchen. Besides an addition to your living space, this will give you a great opportunity to enjoy the natural surroundings of your property.

E. Working with Competent People to Bring Ideas into Reality

You can easily feel overwhelmed, especially if you do not have any prior experience in either design or construction. Thankfully, your ideas can be brought to life, thanks to the professionals. Architectural and interior design pros that cover a barndominium take the modifications you want and work them into a design that is beautiful and useful.

With a builder, you have to make sure you describe what you want. Most builders are used to custom work and can work with you to ensure the end product is what you envision. Other builders would even offer design consultations wherein you can already see whether there is a possibility of altering your floor plan in case construction hasn't started yet.

CONCLUSION

Perhaps the thing that makes the floor designs of barndominiums so intriguing to people who own homes includes the variety they offer: new modern conveniences along with the rustic charm. Many designs are laid out to include open-concept design that accommodates seamless integration between living, kitchen, and dining spaces.

The feeling of freedom and flow that is generated by openness makes it perfect for families or individuals who like their space. Large barn-like structures also facilitate ample room for modification; which allows homeowners to incorporate a wide range of facilities, such as home offices and extended garages.

Cost-effectiveness is a big advantage of barndominium floor plans when compared to conventional house plans. Because of their steel construction and straightforward design, construction can be completed faster in barndominiums, normally in a very short time at lesser costs. As such, they are cost-effective and highly attractive for people who would want to build their dream homes without breaking their budgets. Steel frames are known for their durability, which thus guarantees the lifespan and low maintenance requirements, hence, over the long run, further reducing the expenditures.

Another essential characteristic of barndominium floor designs is the flexibility in combining residential and functional rooms. While many owners use these buildings as residences, they also serve as workshops, hobby rooms, and even business places for their small businesses.

Those owners who would like to easily integrate their work life with their home life will find that these open, spacious designs accommodate such dual-purpose use, and they will therefore be ideal for them. Due to their flexibility in form, it's possible to adapt a barndominium for virtually any lifestyle and set of needs. Accordingly, they also have become one of the most popular choices for people who value practicality just about as much as making a statement with their creativity.

Without further ado, the floor plans of barndominiums reflect the perfect blend of affordable, flexible space in a modern style. Their open designs also make them easy to customize, and since they can be made out of materials that are durable yet inexpensive, they're a great choice for many homeowners.

Whether a large family home, a versatile workspace/life space, or even a uniquely particular architectural style, barndominiums provide an answer that melds the beauty of rustic country living with modern city conveniences. Their growing popularity acts as an indication of the demand for homes to be more functional in use and at least aesthetically beautiful. In turn, these homes are adaptable as much as the structures themselves can allow for a lifestyle that expresses this trait.

Made in United States
Orlando, FL
10 July 2025

62839708R00103